IDEAS TO SAVE YOUR LIFE

THE 15-MINUTE EINSTEIN

SIRIUS

SIRIUS

This edition published in 2017 by Sirius Publishing, a division of
Arcturus Publishing Limited,
26/27 Bickels Yard, 151–153 Bermondsey Street,
London SE1 3HA

Copyright © Arcturus Holdings Limited

All rights reserved. No part of this publication may be reproduced,
stored in a retrieval system, or transmitted, in any form or by any
means, electronic, mechanical, photocopying, recording or otherwise,
without written permission in accordance with the provisions of the
Copyright Act 1956 (as amended). Any person or persons who do any
unauthorised act in relation to this publication may be liable to criminal
prosecution and civil claims for damages.

ISBN: 978-1-78428-454-1
AD005421UK

Printed in China

CONTENTS

INTRODUCTION
Who was Albert Einstein? 4

CHAPTER 1: Does the Universe run like clockwork? 17

CHAPTER 2: What is light? 31

CHAPTER 3: How do light waves travel through space? 45

CHAPTER 4: How did Einstein learn about the quantum? 51

CHAPTER 5: What was Einstein's theory of the photoelectric effect? 59

CHAPTER 6: How did Einstein prove the existence of atoms and molecules? 67

CHAPTER 7: What was Einstein's theory of special relativity? 77

CHAPTER 8: What were Einstein's ideas about time? 90

CHAPTER 9: How did Einstein explain the Lorentz-Fitzgerald contraction? 100

CHAPTER 10: What is spacetime? 109

CHAPTER 11: Why does $E = mc^2$? 121

CHAPTER 12: How did Einstein fit gravity into relativity? 135

CHAPTER 13: What does Einstein say gravity actually is? 150

CHAPTER 14: How did an eclipse prove Einstein was right? 162

CHAPTER 15: If Einstein was right, was Newton wrong? 173

CHAPTER 16: Why didn't Einstein's relativity theory win the Nobel Prize? 182

CHAPTER 17: What was Einstein's greatest blunder? 195

CHAPTER 18: Where does Einstein's relativity theory break down? 208

CHAPTER 19: How did relativity lead to a Big Bang? 221

CHAPTER 20: Does God play dice with the Universe? 236

CHAPTER 21: Einstein versus Bohr – who won? 258

CHAPTER 22: Was Einstein the 'father of the atomic bomb'? 274

CHAPTER 23: Can we find a theory of everything? 288

Who was Albert Einstein?

'Why is it that nobody understands me, and everybody likes me?'

Albert Einstein – from an interview published in the *New York Times*, 12 March 1944.

Albert Einstein was everyone's idea of what a scientist should be like: with his unkempt hair – absent-mindedly puffing on a pipe as he contemplated matters that seemed far beyond the comprehension of ordinary mortals. He was a genius and, beyond doubt, one of the greatest scientists who ever lived. But he had a very human side, too. It is said that when he was an old man in Princeton, New Jersey, children would rush to see him and he would amuse them by wiggling his ears at them.

Albert Einstein was born on 14 March 1879, in Ulm, Germany. He was the first child of Hermann and Pauline Einstein, a Jewish couple. Young Albert was supposedly slow in learning to talk, sometimes to the irritation of his family. 'I very rarely think in words,' he later said. 'A thought comes in and I may try to express it in words afterwards.' In June 1880, the family moved to Munich, where Hermann and his brother Jakob founded an electrical engineering company. Einstein's sister Maja was born in November 1881. On seeing her for the first time, Einstein exclaimed, 'Yes, but where are the wheels?'

When he was aged around four or five, young Albert was ill in bed and his father gave him a compass to play with. He was fascinated by it and the mysterious invisible forces that made the needle swing round. He later said that this had made a deep and lasting impression on him and awakened his curiosity about the world.

Albert liked to work on puzzles and make complex structures with his building set. In 1885, his mother, a skilled pianist, arranged for Albert to have violin lessons. It began a love of music that lasted throughout his life and soon mother and child were playing Mozart duets together.

Einstein started his primary

education at a Catholic school in Munich that same year. He was often at the top of his class. There is a persistently repeated story that Einstein was a poor student of mathematics in school, but it just isn't true. When that was told to Einstein in 1935, he laughed and declared that he had never failed at maths – usually he had been either first or second in his class. 'Before I was 15 I had mastered differential and integral calculus.'

In June 1894, the family moved to Italy, leaving 16-year-old Albert behind in Munich to finish school. Einstein missed his family and grew depressed. He obtained a certificate from his family doctor citing nervous disorders and was released from school. In spring 1895, he travelled to join his family.

Einstein took his entrance exam for the polytechnic in Zurich in October 1895. Though he did well in maths and science, he did not do well enough overall to gain admission, so he attended the Kantonsschule in the town of Aarau to gain the certificate he needed for the polytechnic. In January 1896, Einstein renounced his German citizenship and that autumn, having passed his exams, he registered as a resident of Zurich and became a student at the polytechnic with the aim of becoming a teacher in maths and physics. He then formally applied for Swiss citizenship, which was granted on 21 February 1901.

After gaining his diploma degree in 1900, Einstein began to look for work, applying, without success, for assistantships at the polytechnic and other universities. Finally, in May 1901, he found a temporary job as a substitute teacher for two months at a high school in Winterthur. This was followed by another temporary position at a private school in Schaffhausen. While he was there, he wrote his doctoral dissertation on the kinetic theory of gases, but it wasn't accepted.

In 1902, Einstein moved to the Swiss capital, Bern, hoping to get a job at the patent office. In order to make a living in the meantime, he gave private lessons in maths and physics.

In January 1902, Einstein had a daughter, Lieserl, with Mileva Maric, who had been a fellow student at the polytechnic in Zurich. The existence of Einstein's illegitimate child only came to light when private letters mentioning her were published in 1986. Einstein apparently told neither his family nor his friends about the child, and, it seems, never even laid eyes on her (she was born at Maric's family home in Hungary). Nothing is known about the life of Einstein's daughter; not once did he publicly acknowledge her existence. It is likely that she was either given up for adoption, or died in infancy.

On 16 June 1902, Albert Einstein became technical expert third class at the Bern patent office on a trial basis. At the end of 1902, Einstein's father became gravely ill in Milan. Einstein travelled from Bern to Milan to be with his father, who finally, on his death bed,

PATENTS

consented to his son's marriage to Mileva. On 6 January 1903, Einstein married Mileva Maric, much to the disapproval of both families. In May 1904, Einstein's first son, Hans Albert, was born, followed in July 1910 by his second son, Eduard.

Einstein enjoyed his job at the patent office. He took his work seriously but still managed to find enough time and energy to continue his physics research. Reminiscing with his friend Michele Besso years later, Einstein wrote about: '...these days in that temporal monastery, where I hatched my most beautiful ideas and where we spent such pleasant time together.'

In April 1905, Einstein submitted his doctoral thesis, 'A New Determination of Molecular Dimensions', to the university in Zurich. It was accepted in July. It was the start of a remarkable outpouring of ideas – no one before or since has changed science in such a profound way in as short a time as Albert Einstein managed in 1905.

The list of Einstein's achievements in his 'miracle year' of 1905 makes for an impressive body of work:

1. **'On A Heuristic Point of View Concerning the Production and Transformation of Light'**, completed 17 March.
   (This is the paper on light quanta and the photoelectric effect that eventually won him the Nobel Prize in physics, and was produced before he wrote his PhD thesis.)

2. **'A New Determination of Molecular Dimensions'**, completed 30 April.
   (His doctoral thesis, which became the paper of Einstein's that was most often quoted in modern scientific literature.)

3. **'On the Motion Required by the Molecular Kinetic Theory of Heat of Particles Suspended in Fluids at Rest'**, submitted 11 May.
   (This was Einstein's 'Brownian motion' paper and followed on from his thesis work.)

4. **'On the Electrodynamics of Moving Bodies'**, submitted 30 June.
   (This was the first paper on special relativity.)

5. **'Does the Inertia of a Body Depend upon its Energy-Content?'** submitted 27 September.
   (The second paper on special relativity, containing the famous $E = mc^2$ equation.)

6. **'On the Theory of Brownian Motion'**, submitted 19 December.
   (A follow up to his earlier paper on 'Brownian motion'.)

In April 1906, Einstein was promoted to technical expert second class at the patent office. His first application, in 1907, for a professorial position at the university of Bern was turned down. In early 1908, however, he was successful and gave his first lecture at the end of that year. Deciding that he wanted to devote his life to science, Einstein resigned his position at the patent office in October 1909 and that same month began work as an adjunct professor of theoretical physics at the university of Zurich. He was offered a chair at the German university in Prague in 1911, which he accepted, but returned to Switzerland after a year to take up a professorial position at the Zurich polytechnic.

Impressed by what Einstein had achieved, the physicist Max Planck (1858–1947) offered him a professorial position without teaching responsibilities at Berlin university, to make him a member of the Prussian Academy of Sciences, and head of the planned Kaiser-Wilhelm-Institute of Physics. This was too enticing an offer to pass by and Einstein accepted enthusiastically, taking his family to Berlin in April 1914.

Unfortunately, Einstein's marriage did not go as well as his career. In July 1914, after only a few months in Berlin, Mileva took the children back to Zurich. The couple eventually divorced in February 1919. From 1917 until 1920, Einstein suffered ill health and was generally very weak. During this time he was nursed by his cousin Elsa

Loewenthal, who he married on 2 June 1919. Elsa had two daughters, Ilse and Margot, from her first marriage. Charlie Chaplin, who met Elsa in 1931, described her as 'a square-framed woman with abundant vitality; she frankly enjoyed being the wife of the great man and made no attempt to hide the fact; her enthusiasm was endearing.'

Between 1909 and 1916, Einstein was hard at work on the General Theory of Relativity, which he eventually published in March 1916 as 'The Foundation of the General Theory of Relativity'. One consequence of the theory was its prediction that the light from a distant star would be bent by the gravity field of a massive body, such as the Sun. This was confirmed in 1919 by the British scientist Arthur Eddington who observed Einstein's predicted bending of starlight near the Sun during a total eclipse (see page 164). J. J. Thomson, president of the Royal Society, declared it '...the most important result related to the theory of gravitation since the days of Newton...[it] is among the greatest achievements of human thinking.'

In the early days of the First World War, Einstein spoke out publicly for pacifism. It was a lifelong concern of his. He met a hostile response; the chief of staff of the Berlin military district advocated taking pacifists, including Einstein, off the streets. The General Theory of Relativity had placed Einstein very much on the public stage. Invitations and honours arrived from all over the world. But there was also a downside to this newfound fame. Einstein and his theory were

Who was Albert Einstein? | 13

subjected to anti-Semitic abuse. Even some German Nobel laureates were hostile towards Einstein and demanded a 'German physics'.

Einstein was greatly affected by the unrest in Germany in the early 1920s. In 1922, he departed with Elsa for a five-month trip abroad. 'I very much welcomed the opportunity of a long absence from Germany,' he said, 'which took me away from temporarily increased danger.' It was while he was on this trip that he received word that he had been awarded the Nobel Prize.

From 1920 onwards, Einstein was attempting to formulate a unified field theory, one that would unite gravity with electrodynamics; it was a quest that would occupy him until his death, and one he never succeeded in fulfilling. Also at this time, Niels Bohr, Louis de Broglie, Werner Heisenberg, Wolfgang Pauli and other physicists were setting the foundations of the new physics of quantum mechanics. Einstein was unable to accept the theories of quantum mechanics and was constantly challenging it. Today, the tenets of quantum mechanics are as widely accepted as are Einstein's own theories, though science is still vexed as to how the two views can be united.

In December 1932, Einstein and his wife went on a lecture tour of the United States. The political situation in Germany had definitely taken a turn for the worse. In the 1932 elections, the Nazis became the strongest political party and in January 1933 Hitler seized power. Einstein would never return to Germany again. In May 1935, Einstein and Elsa sailed to Bermuda; it was Einstein's last trip outside the United States. Not long afterwards, Elsa became very ill and died of heart disease on 20 December 1936.

In 1939, Einstein's sister, Maja, was forced to flee Italy to escape Mussolini's fascists and came to live with her brother in Princeton, New Jersey. With Europe on the brink of war, Einstein was convinced that scientists in Germany were working on the technology that would lead to an atomic bomb. On 2 August 1939, Einstein signed a letter to US President Franklin D. Roosevelt, drawing his attention to the danger and urging the United States to embark on its own atomic bomb programme.

Einstein became an American citizen on 1 October 1940, though he also retained his Swiss citizenship. In October 1946, Einstein wrote an open letter to the general assembly of the United Nations, urging the formation of a world government; this was an enduring belief of Einstein's and something that he saw as the only way to ensure lasting peace.

In August 1948, Einstein's first wife, Mileva Maric, died in Zurich. In December of that year, Einstein himself had to undergo abdominal surgery. In March 1950, he signed his will making his secretary, Helen Dukas, and Dr Otto Nathan his joint executors. Helen Dukas had become Einstein's secretary in April 1928 and, following the death of his first wife, Elsa, in December 1936, also became his housekeeper.

She remained with Einstein until his death. After he died, Dukas devoted herself to the task of sorting and cataloguing Einstein's papers. Largely thanks to her work, these documents can now be found in the Albert Einstein Archive of the Hebrew University in Jerusalem. In November 1952, Einstein was offered the presidency of Israel but he declined it.

On 11 April 1955, Einstein sent a letter to the philosopher Bertrand Russell in which he agreed to be a signatory to a manifesto urging all nations to renounce nuclear weapons. That same week, in an unfinished manuscript, Einstein set down the last phrase he would ever write: 'Political passions, aroused everywhere, demand their victims.'

On 15 April 1955, Einstein was transported to hospital in severe pain. A few days later, on 18 April, he died. He was 76. The diagnosis was a ruptured aneurysm of his abdominal aorta. In accordance with his wishes, his remains were cremated that same day and the ashes scattered two weeks later at an undisclosed location.

To the end of his life, Einstein retained his sense of curiosity and wonder at the Universe around him. He wrote to a friend: 'People like you and me never grow old. We never cease to stand like curious children before the great mystery into which we were born.'

Chapter 1
Does the Universe run like clockwork?

People have pondered the workings of the Universe for as long as there have been people to ponder. Here are some of the ideas they had.

The Music of the Spheres

It was the great thinker Plato, born around 427BC, who declared that the heavens were perfect and the stars and planets moved in 'perfect curves on perfect solids'. He believed that these spheres produced music as they turned – an idea that would persist for many centuries to come.

The trouble was that the celestial spheres simply didn't fit the evidence of the skywatchers' eyes.

There were a handful of stars that behaved oddly. They appeared to move against the background of the other 'fixed' stars, sometimes even looping backwards in their paths before continuing on their course. These peculiar stars were called 'asteres planetai' by the Greeks, meaning 'wandering stars'. We call them 'planets'.

The ancient Greeks concocted all sorts of complex schemes to explain planetary motion involving spheres moving within spheres within yet more spheres, all rotating in slightly different directions. Around AD100, the astronomer Ptolemy set out a map showing an Earth-centred Universe of nested spheres – an idea that would remain largely unchallenged for 1,400 years. It lasted so long because it worked. Ptolemy's system gave accurate predictions of where the planets would be found at any given time.

Celestial Revolutions

In 1543, astronomy was awakened from its Ptolemaic slumbers by the arrival of a remarkable book called *De revolutionibus orbium coelestium* (*On the Revolutions of the Celestial Spheres*). In 1507, the Polish astronomer and mathematician Nicolaus Copernicus had much the same idea that had occurred to Aristarchus 1,800 years previously. If he assumed that the Sun was at the centre of the Universe, and the Earth and planets orbited around it, some of the puzzles of planetary motion could be explained. Mars, Jupiter and Saturn were further away from the Sun than the Earth which, moving faster in its smaller orbit, sometimes overtook them, making them look, from our point of view, as if they were going in reverse.

Take a moment to imagine hearing for the first time that the Earth wasn't the centre of all creation as you had previously

> **AHEAD OF HIS TIME**
>
> Around 260 BC, the astronomer Aristarchus declared that it was the Sun that was the centre of the Universe, not the Earth. This, he said, explained the movements of the planets without resorting to spherical shenanigans. The stars were infinitely far away and only appeared to move because the Earth rotated beneath them. These prescient notions were deemed too far-fetched by his contemporaries and so largely ignored.

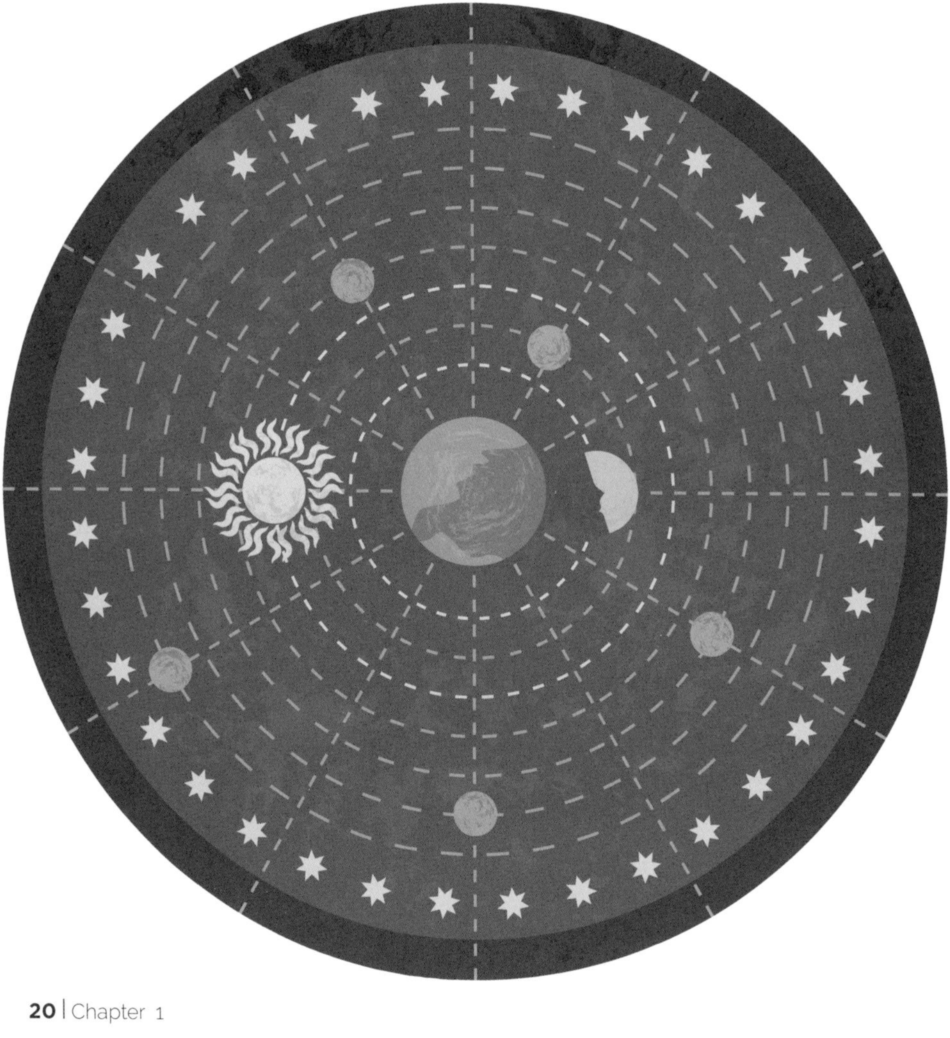

believed. Perhaps unsurprisingly, this was not a popular suggestion, particularly with the powerful Church. No one at that time made an enemy of the Church as the consequences of doing so could be somewhat grim. *Revolutions* was published with an introduction (added without Copernicus' approval) that these revolutionary ideas need not be considered as true. In 1616, *On the Revolutions of the Celestial Spheres* was placed on the Catholic Church's list of banned books. It stayed there until 1835.

Word of the Copernican model spread slowly. Copernicus still believed that the Universe was formed of perfect spheres, just no longer centred on the Earth. But, at the beginning of the 17th century, the German astronomer Johannes Kepler's painstaking observations led him to a sensational conclusion. The paths of the planets were not perfect circles, they were flattened circles, or ellipses. After Galileo's discovery of the moons of Jupiter, Kepler found that they, too, moved in elliptical paths around the giant planet.

Kepler set out his three laws of planetary motion that described how the planets moved, but not why they moved. He tried to work out what force might be responsible for the planets moving as they did. He thought that magnetism might be involved and that the Sun must have something to do with it but couldn't come up with a satisfactory explanation. That would come 50 years later, from Isaac Newton and his ideas about gravity.

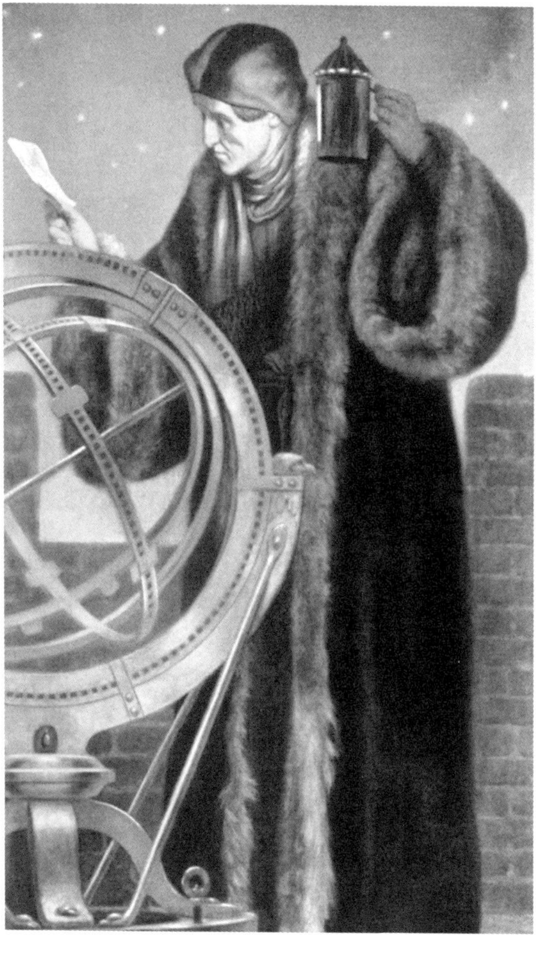

Enter Newton

Until Einstein came along, our understanding of the laws that govern the movement of objects through space was founded on the work of the scientist Isaac Newton (1643–1727). The story of Newton in the apple orchard is an often repeated one, but it really was a singular mind that asked the question: 'Why doesn't the Moon fall to the Earth like the apple does?' And it took real genius to conclude that the Moon actually is falling.

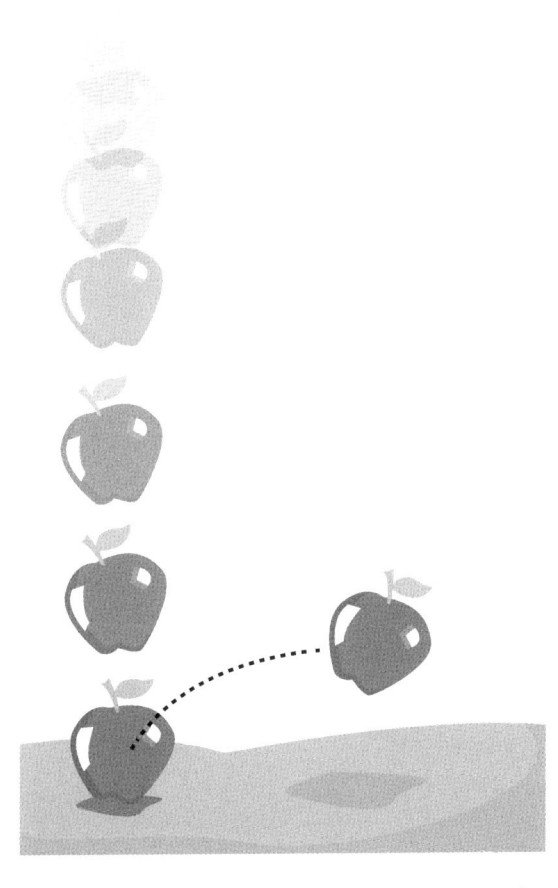

The Universal Force

Newton knew that any explanation he came up with to explain the motion of both the apple and the Moon would have to explain Kepler's findings too. In 1687, Newton produced what is reckoned by many to be a contender for the greatest work of science ever written. The *Philosophiae Naturalis Principia Mathematica* (*Mathematical Principles of Natural Philosophy*), often referred to simply as the *Principia*, set out Newton's vision of a Universe where all events took place against a backdrop of infinite space and smoothly flowing time.

Building on experiments carried out on moving objects by Galileo and on Kepler's observations of the planets, Newton set out his three laws of motion and his theory of gravity.

> **NEWTON'S LAWS OF MOTION**
> 1: An object will remain at rest or continue to move in the same direction and at the same speed unless acted on by a force.
> 2: A force acting on an object will cause it to move in the direction of that force. The magnitude of the change in the speed or direction of the object is dependent on the size of the force and the mass of the object.
> 3: For every action there is an equal and opposite reaction. If one object exerts a force on another an equal and opposite force is exerted by the second object on the first.

Does the Universe run like clockwork? | 25

Newton determined that between any two objects there is always a gravitational force that attracts them to each other. The strength of the force depends on the masses of the objects and on the distance between them. Gravity obeys an inverse square law, which means that the magnitude of the force decreases by the square of the distance. Therefore, if you double the distance between two objects the force that draws them together reduces to just a quarter of what it was. At five times the distance the force is reduced to a 25th of what it was.

With three simple laws of motion and one law of gravity, Newton, it seemed, could explain the movement of everything in the Universe. Newton's laws provided an explanation for Kepler's laws of planetary motion and for the fall of the apple. Newton derived his laws from three fundamental quantities that underpin all of science – time, mass and distance. Knowing the time an object takes to travel a set distance enables you to calculate its velocity (speed and direction). Mass tells you how much matter that object contains and therefore the amount of force you'll need to move it. Multiply the mass by the velocity and you'll get the object's momentum, which will tell you how hard it's going to be to stop it once it is moving. Einstein was to show that all three of these quantities were relative.

Does the Universe run like clockwork? | **27**

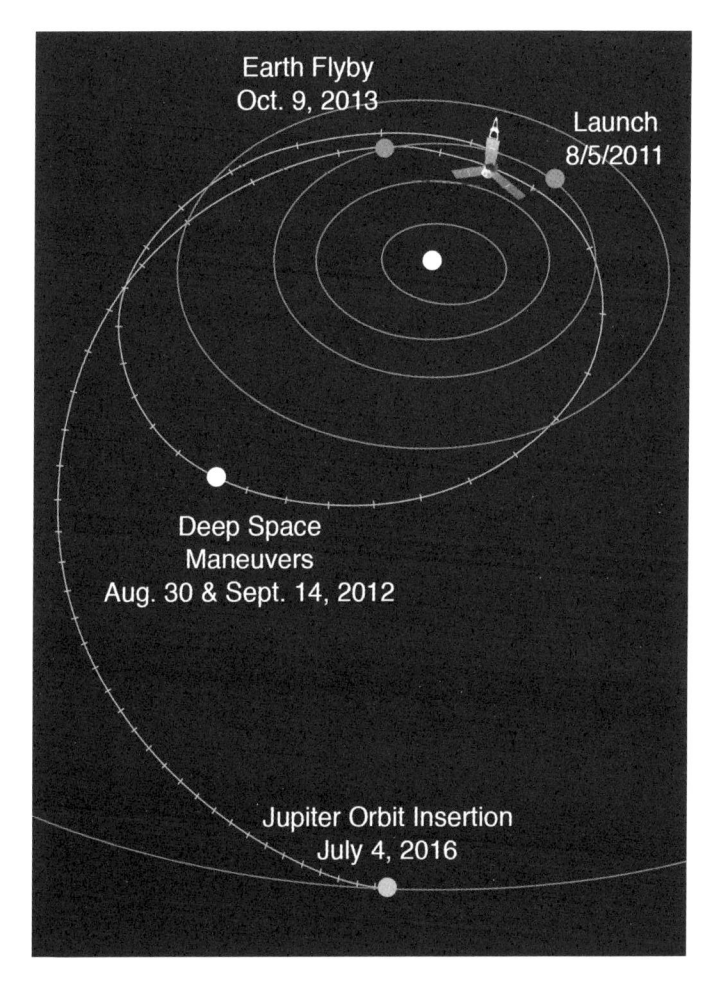

Absolute time and space

According to Newton, time and space were absolute; they were the stage upon which the drama of the Universe unfolded, remaining unchanged by events. Newton thought of our everyday measures of the passing of time – the hour, the month, the year – as simply common time. These were useful, of course, but in no way were they to be confused with 'true', or 'absolute', time, as Newton called it. Absolute time, he believed, was completely separate from space and independent of events. Absolute time ticked along at the same steady pace throughout the Universe. One second for you should be exactly the same as one second for me, wherever we were in the Universe and whatever we were doing.

Newton also believed in the idea of absolute space. He thought it should be possible to state the absolute position of an object in absolute space, almost as if you could cover the Universe in three-dimensional graph paper and plot the positions of everything in it. But it is just as impossible to say what absolute space was, as it is to define absolute time.

Newton's laws were unchallenged for more than 200 years. For everyday purposes, they are still an excellent way of calculating the movement of an object and how it is affected by gravity. But Newton had failed to explain what caused gravity. When Einstein came along he offered up a suggestion that might have astonished even Newton.

Does the Universe run like clockwork? | **29**

COSMIC CANNONBALLS

Two forces govern a cannonball's path – gravity and the force that propelled it from the cannon. The result of those two forces acting on it is that it follows a curved path back to earth. Imagine that the cannon produced sufficient force that the curved path of the cannonball now matched the curvature of the Earth. It would now travel right around the Earth, always falling around the planet but never reaching the ground. (Let's assume for the sake of the argument that there's no air resistance to slow it down.) The cannonball is now a satellite in orbit. This is exactly the principle that puts real satellites into orbit, except with powerful rockets rather than cannon providing the forward motion. The Moon is like a cosmic cannonball, perpetually falling around the Earth in its orbit.

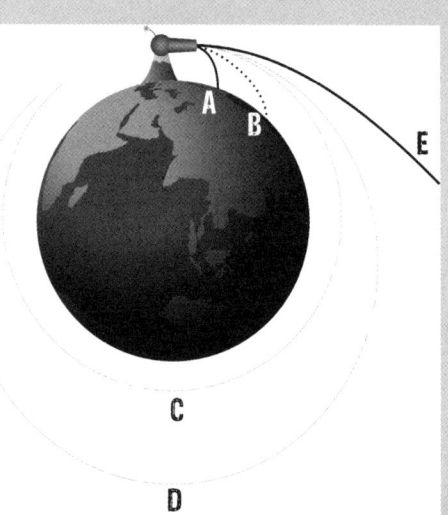

Chapter 2
What is light?

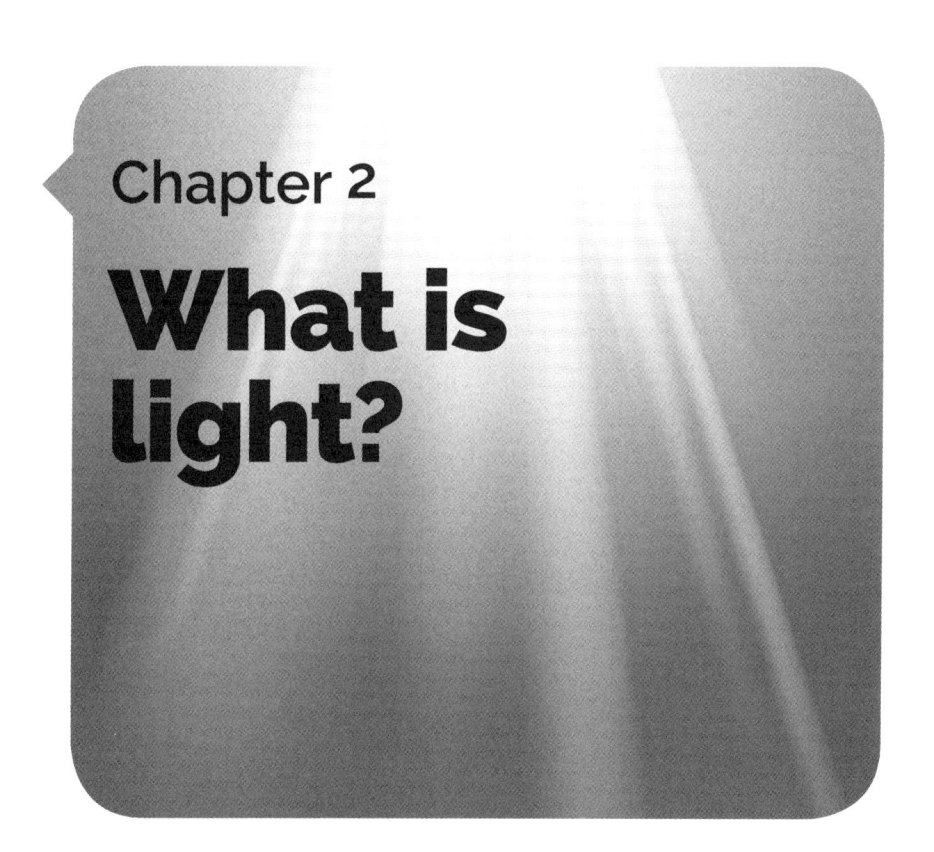

The nature of light underpins much of Albert Einstein's work, but what is it actually?

The nature of light is at the heart of Einstein's work. For centuries, people have tried to explain the various phenomena associated with light. The ancient Greeks certainly had a good go at it. The 6th century BC Greek philosopher Pythagoras thought sight was like a very delicate sense of touch, the eyes producing invisible rays with which we sense objects. Another Greek thinker, Democritus, believed that objects continually emitted images of themselves, which we sensed.

The obvious drawback to both these ideas was this: why then can't we see well at night? Plato put forward the idea that the inner light from the eyes had to mix with the light from the Sun before we could see anything. Aristotle suggested that we could only see things if they were illuminated but this notion was rejected as being just too simple!

Waves or particles?

Whatever they believed its origins to be, the Greeks also had opposing views as to the nature of light. The first was that light was a disturbance in the ether, an invisible, undetectable substance that filled space. This was another of Aristotle's ideas. He saw light as a wave that travels through the ether like an ocean wave travels across the water. Another view held that light was a stream of tiny particles that were too small and fast-moving to be perceived individually. Plato and Aristotle both opposed the particle theory and so, for the next 2,000 years or so, it was generally accepted that light travelled in waves.

Let there be light…

The Arabian physicist Alhazen (965–1038) finally put to rest the idea that beams of light emanated from the eyes. He established once and for all that we see things either because they reflect light from a source of illumination, or they are themselves a source of illumination, whether it be a candle or the Sun.

The English scholar Robert Grosseteste (c.1168–1253) read Alhazen's work and carried out some experiments of his own. He believed that the entire Universe had been formed from light. Light was the

> **BEND IT LIKE BACON**
>
> Roger Bacon (c.1220–92) was an English monk and a pupil of Grosseteste's who shared his teacher's enthusiasm for studying light. By some accounts, Bacon was the first modern scientist, placing great emphasis on the importance of carrying out experiments. Some of these experiments involved bending and focusing light by passing it through a lens. Bacon was one of the first to suggest spectacles for people with poor eyesight.

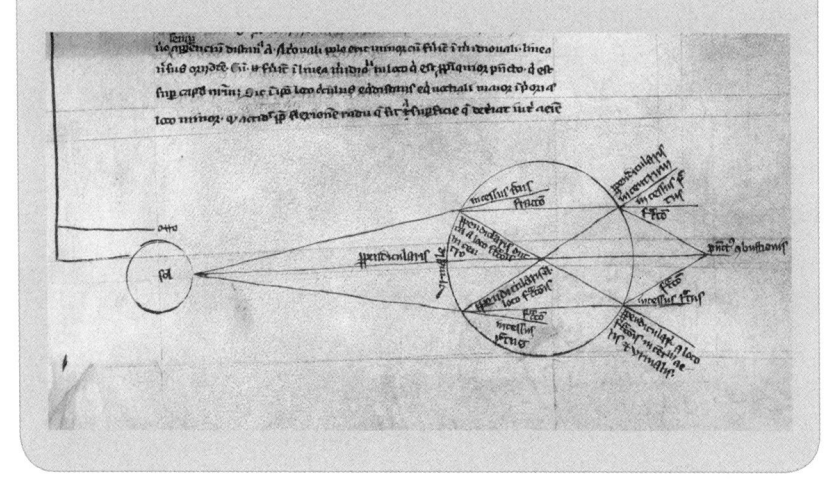

34 | Chapter 2

first of all things to be created, expanding out from a single point into a sphere that contained all other things within itself. This was a startlingly sophisticated notion with obvious parallels with our current thinking of how the Universe formed.

Particles or waves?

In addition to his work with motion and gravity, Isaac Newton was also fascinated by light. He carried out many experiments and had his own ideas about the nature of light. He demonstrated that white light could be split into a rainbow of colours – a spectrum – by passing it through a prism. He noticed that light travels in straight lines and that shadows have sharp edges. It seemed obvious to Newton that light was a stream of particles, not a wave.

Thomas Young (1773–1829) was a man of such formidable intellect that his fellow students at Cambridge University dubbed him 'Phenomenon'. He had different ideas on light.

Young decided to tackle the wave/particle problem by experiment. He theorized that if the wavelength of light was sufficiently short then it would appear to travel in straight lines, as if it were a stream of particles. In 1803, he carried an experiment that was beautiful in its elegance and simplicity.

He began by making a small hole in a window blind, which gave him a point source of illumination. Next, he took a piece of board and made two pinholes in it, placed close together. He positioned his board so that the light coming through the hole in the window blind would pass through the pinholes and on to a screen. If Newton was right, and the light was a stream of particles, then there would be two points of light on the screen where the particles travelled through the pinholes. So what did Young actually see?

Waving goodbye to particles... for now

Two years earlier, in 1801, Young had described an effect he called 'interference'. If two waves meet they don't bounce off each other like colliding snooker balls; instead, they appear to pass straight through each other. Watch raindrops falling on to a pond and see how the ripples spread and meet and keep on going as they cross each other.

Where the waves cross they combine with each other. If the peak of one wave meets the peak of another they are added together to make a higher peak; two troughs make a deeper trough and a trough and a peak cancel each other out. The result was an interference pattern that showed where the waves were adding and cancelling. And this was what Young saw on his screen.

Rather than two discrete points of light, he saw a series of curved, coloured bands separated by dark lines, exactly as would be expected if light were a wave. Unfortunately, it was not really the done thing to contradict the great Newton and Young's findings were not well received.

But still the question remained. What actually was light? And if it really was a wave, how did it travel through space? The beginnings of an answer came from the study of a seemingly unrelated force – electricity.

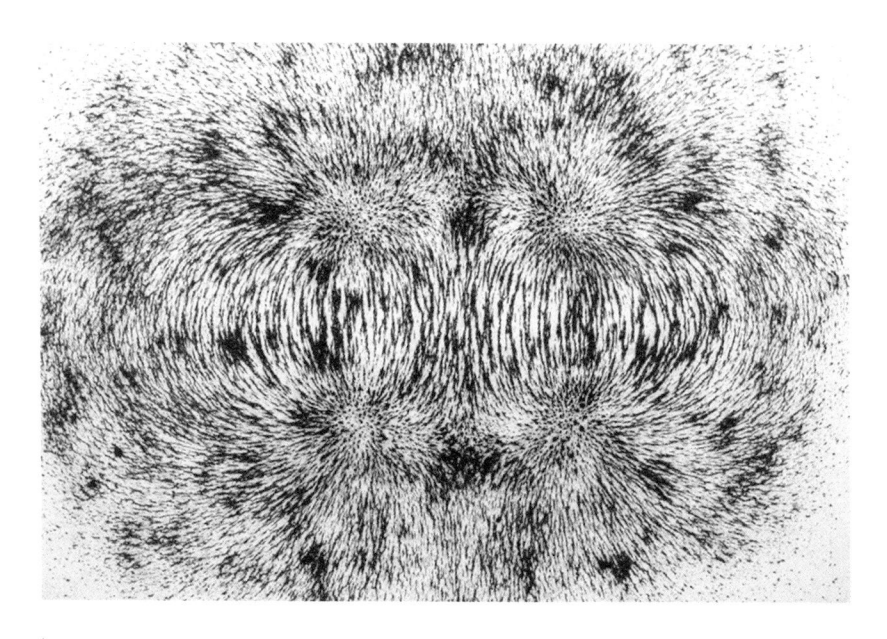

Secrets of electromagnetism

As the 19th century progressed, scientists were gaining more and more knowledge about electricity. Among their discoveries was the intimate relationship between electricity and magnetism.

In 1820, Danish physicist Hans Christian Øersted discovered that a wire carrying an electric current would deflect the needle of a compass. French scientist André-Marie Ampère (who gave his name to the amp) experimented further. He found that if he placed two wires, each carrying an electric current, close to each other one of two things happened. If the current was flowing in the same direction in each wire the wires were pushed apart. If the current was flowing in opposite directions the wires were drawn together. They were, in other words, behaving just like magnets.

Electromagnetic induction

In the course of hundreds of experiments carried out at the beginning of the 19th century, the English scientist Michael Faraday (1791–1867) discovered that just as an electric current produced magnetism so a magnet moving through a coil of wire generated electricity. The magnet had to be moving – if it was stationary nothing happened. Electric currents make magnetic fields and moving magnets produce electric currents, a process that is now called electromagnetic induction. That magnetism and electricity were related on some fundamental level could no longer be in doubt. Our world would be a very different place without this discovery. It is the principle behind all of the electricity generated in the world's power stations and what drives all of the countless electric motors we use.

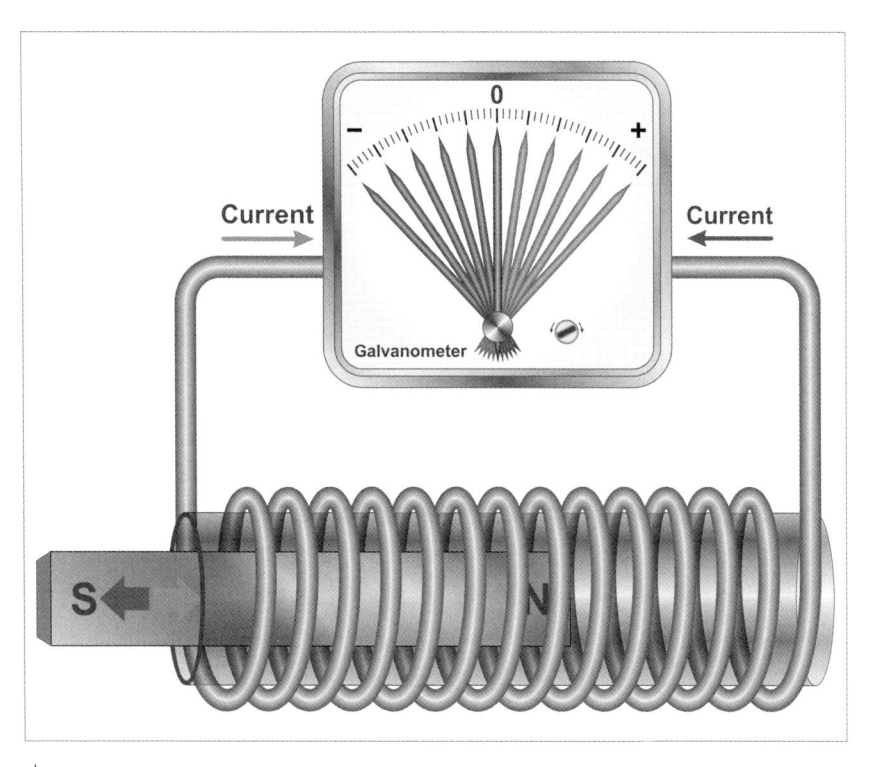

The Faraday effect

Faraday was convinced that there was a link between electricity, magnetism and light. In 1845, he carried out an experiment in the basement of the Royal Institution in London and found that he could affect the polarization of a beam of light using an electromagnet, so showing that light did indeed have magnetic properties. Faraday wrote in his notebook: 'I have at last succeeded in … magnetizing a ray of light'. This demonstration of what is now called the 'Faraday effect' was an important stepping stone in Faraday's development of the field theory of electromagnetism, which in turn would influence the work of the physicist James Clerk Maxwell and later Albert Einstein.

Fields and forces

Faraday wanted to explain how a magnet could induce an electric current in a wire without coming into physical contact with it, or an electric current make a compass needle move. To do so, he came up with the idea of an electromagnetic field. He saw this as lines of force, which he called 'flux lines', stretching invisibly through all of space. It's easy enough to make these lines visible – simply place a magnet under a sheet of paper and scatter some iron filings on it. The patterns formed by the filings reveal the magnetic lines of force. According to Faraday's field theory, the magnet was not the centre of the magnetic force but rather it concentrated the force through itself. The magnetic force wasn't in the magnet but in a magnetic field in the space surrounding it. The magnetic lines of force were concentrated around magnets, rather than the magnets actually creating the fields.

Towards a deeper understanding

Some 20 years after Faraday proposed his field theory, the idea was taken up by Scottish physicist and mathematician James Clerk Maxwell (1831–79), who set out to express Faraday's ideas mathematically. Albert Einstein himself would later describe Maxwell's work on electromagnetism as 'the most profound and most fruitful that physics has experienced since Newton'.

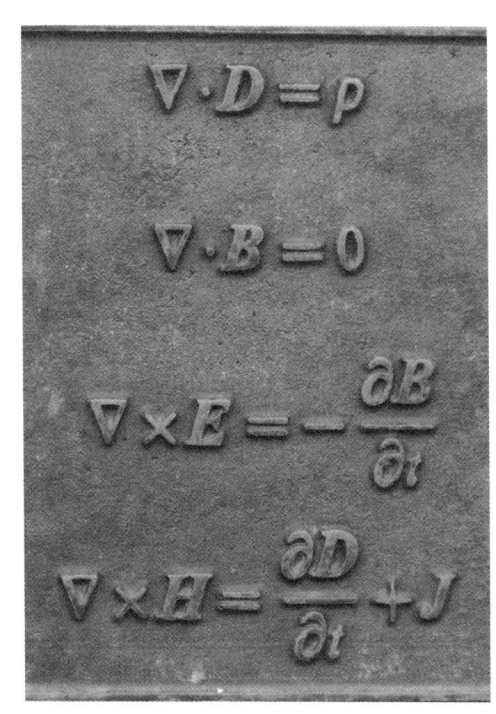

$$\nabla \cdot D = \rho$$

$$\nabla \cdot B = 0$$

$$\nabla \times E = -\frac{\partial B}{\partial t}$$

$$\nabla \times H = \frac{\partial D}{\partial t} + J$$

In just four short equations, Maxwell succeeded in describing all of the electric and magnetic phenomena that had been observed and recorded by Faraday and other researchers. Maxwell's equations described all of the different aspects and behaviour of both forces and provided accurate predictions for future experiments. When Einstein got busy upsetting everyone's ideas about the way the Universe worked, Maxwell's equations stood up to the challenge and came through unscathed.

From his equations, Maxwell proposed the existence of electromagnetic waves. You might try to imagine an electromagnetic wave as being like two waves travelling in the same direction but at right angles to each other. One of these waves is an oscillating magnetic field, the other an electric field doing the same thing. The two fields keep in step with each other as the wave travels along. Maxwell saw that electricity and magnetism were always bound together and that it was impossible to have one without the other.

Maxwell used his equations to calculate the velocity of an electromagnetic wave. The answer he obtained was 299,792,458 metres per second (m/s). This was in agreement with what experiments had shown to be the velocity of light. Maxwell thought this couldn't possibly be a coincidence and had no hesitation in declaring that light itself was an electromagnetic wave.

THE ELECTROMAGNETIC SPECTRUM

Maxwell predicted that there should be a whole range, or spectrum, of electromagnetic waves, and so it proved. Infrared and ultraviolet light, invisible to human eyes, had already been discovered at either end of the visible spectrum and scientists had demonstrated that they had the same wave-like properties as visible light. After Maxwell's death, the discovery of long wavelength radio waves and very short wavelength X-rays and gamma rays extended the spectrum further.

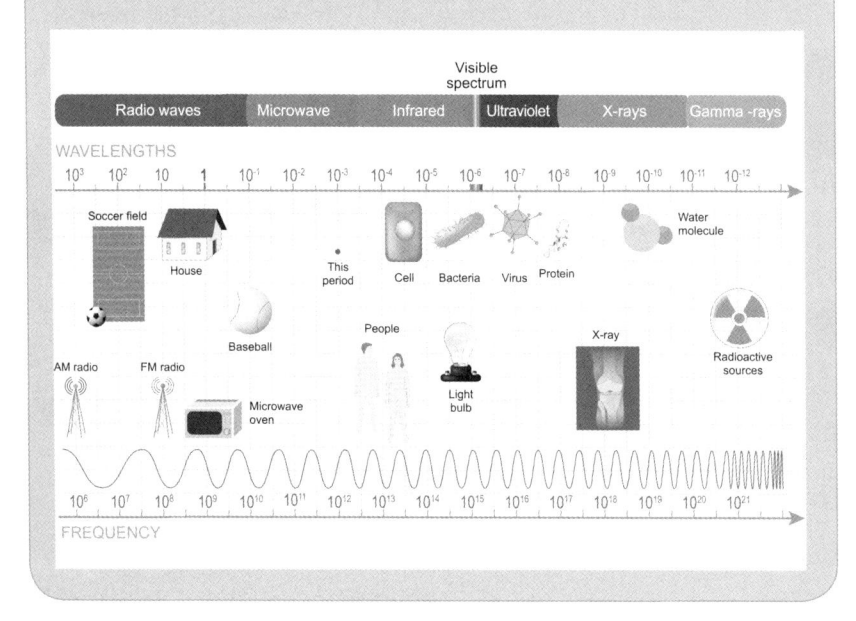

Chapter 3
How do light waves travel through space?

Michelson and Morley failed to find the ether and Lorenz and FitzGerald had a "shrinking" feeling – time for Einstein!

If light is a wave, as Maxwell suggests, how does it travel through the vacuum of space? Waves, after all, need some sort of medium to carry them. It's easy enough to make waves in the water of a swimming pool by waving your arms up and down; if you clap your hands you can send a sound wave through the air. But how does the light from the Sun, say, get from there to here? It was the view of Maxwell and his contemporaries that light must also travel through a medium. They called it 'ether'.

Ether, or...?

This ether was an odd substance. To all intents and purposes it appeared to be undetectable. It seemingly offered no resistance to planets or any other objects passing through it. Light apparently passed through it undiminished and failed to illuminate the ether in any way. Yet it must fill all of space since the light from the stars reaches us from every direction. How could this ghostly substance that filled the Universe be detected?

The ingenious experimenters of the 19th century set out to find a way to pin down the elusive ether. Two particularly dedicated investigators were the American scientists Albert Michelson and Edward Morley, who embarked upon a series of precise experiments to demonstrate the effects of ether on the light that passed through it.

Against the wind

As the Earth moved in its orbit around the Sun, the flow of the ether across the Earth's surface could, it was believed, produce an 'ether wind'. A beam of light travelling through the ether should travel faster if it was moving with the wind and slower if it was going against it. The aim of Michelson and Morley's crucial experiment, carried out in 1887, was to measure the speed of light in different directions and so determine the speed of the ether relative to the Earth.

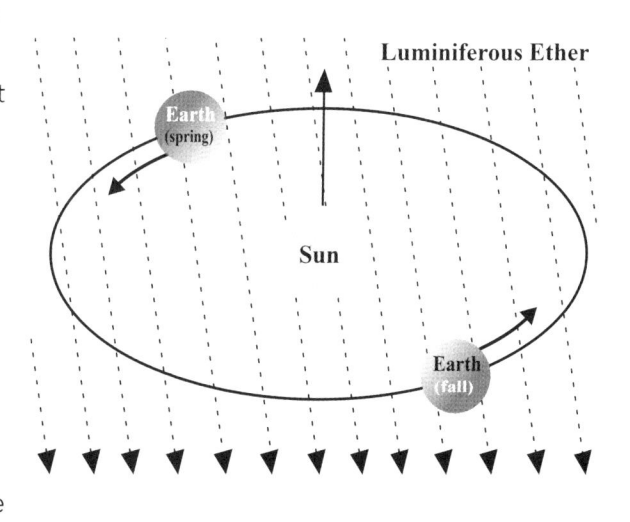

To carry out the measurements, Michelson designed a device called an interferometer. This sent the beam from a light source through a half-silvered mirror; this split the light into two beams travelling at right angles to one another. The beams were then reflected back to the middle by two more mirrors. The beams recombined, producing an interference pattern that could be observed through an eyepiece. Any variation in the time it took for the beams to travel between the mirrors would be seen as a shift in the interference pattern. The interferometer floated in a trough of mercury, allowing it to be rotated slowly. If the ether theory was correct, the speed of the light beams would change as their direction changed in relation to the direction of the Earth's orbit.

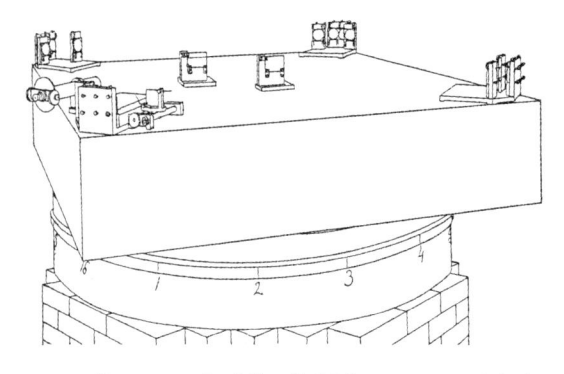

Michelson and Morley discovered that it made not one bit of difference how they rotated the apparatus. Nor did it matter at what time of day they took their measurements. The speed of the light beams was always the same. It seemed that the ether didn't exist after all.

Searching for an explanation

The world of physics was shaken by the failure to find evidence for ether. No one doubted the reliability of the experiment, but there was reluctance to accept its result. Physicists looked for a way to explain the findings and yet still hold on to the existence of the ether. Michelson himself was perplexed by the results. He repeated the experiment, even trying it on a mountaintop, but the speed of light remained the same – there was not the slightest indication that the ether existed. Michelson even wondered if this was because the ether stuck to the Earth and was dragged around with it.

The Lorentz–FitzGerald contraction

Dutch physicist Hendrick Lorentz and Irish physicist George FitzGerald, working independently, came up with the same solution to the problem. In 1889, FitzGerald published a short paper of less than half a page in which he proposed that the results of the Michelson–Morley experiment could be explained only if objects reduced in length as they travelled through the ether.

Unaware of FitzGerald's idea, Lorentz put forward an almost identical proposal in 1892. When it was pointed out to Lorentz that FitzGerald had published a similar theory, he took every opportunity to acknowledge that FitzGerald had proposed the idea first.

The reduction in length suggested was infinitesimal, amounting to just a couple of centimetres for an object the size of the Earth, but it would be enough to explain Michelson and Morley's results. This may seem like a bit of a fudge. Where was the proof that objects shrank?

A few years later, a clerk working in a Swiss patent office would show that the whole idea of the ether was simply unnecessary. All you had to do, Albert Einstein would suggest, is to give up the notion of absolute time.

HOW DO ELECTROMAGNETIC WAVES TRAVEL THROUGH SPACE?

An electromagnetic wave is produced by changes in the electric and magnetic fields. As demonstrated by Faraday and others, a changing electric field produces a changing magnetic field and a changing magnetic field produces a changing electric field. An electromagnetic wave is self-propagating and does not need a medium to travel through. Unlike a water wave displacing water molecules as it passes through them, nothing is displaced in space by an electromagnetic wave. You might think of an electromagnetic wave as an energy-carrying disturbance travelling invisibly through space until it interacts with matter.

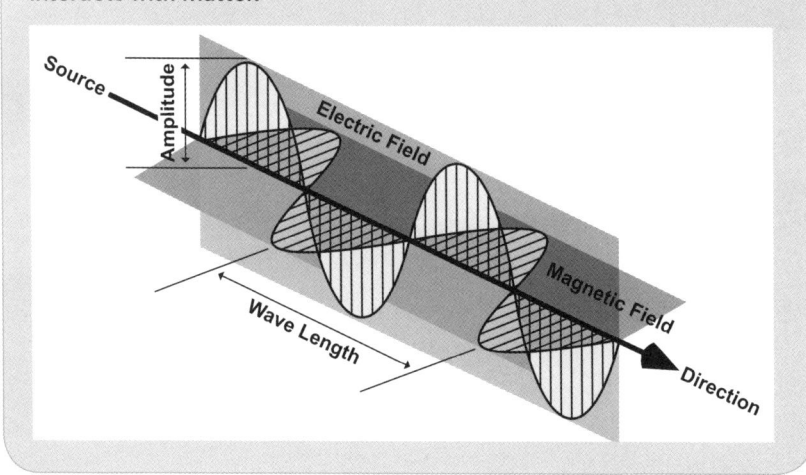

Chapter 4

How did Einstein learn about the quantum?

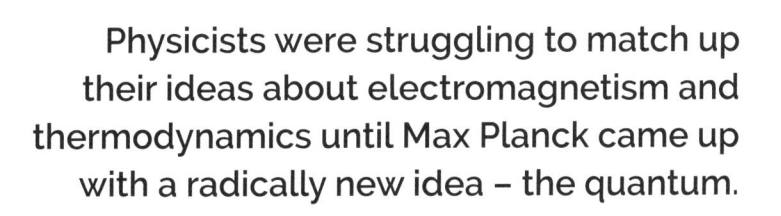

Physicists were struggling to match up their ideas about electromagnetism and thermodynamics until Max Planck came up with a radically new idea – the quantum.

If you heat up a metal rod sufficiently it begins to glow – its gets red hot. Heat it more and it glows yellow and eventually white hot, giving off light at all wavelengths of the spectrum. But why should this be? Why does increasing the temperature cause electromagnetic waves to be produced?

Blackbody radiation

The fact is that every object gives off electromagnetic radiation all the time. Just how much radiation, called 'blackbody radiation', it gives off depends on the object's temperature. A blackbody is simply something that absorbs the electromagnetic radiation that hits it and then radiates it back out again, mostly in the form of infrared radiation, which we detect as heat. The hotter an object is the more energetic the blackbody radiation it emits. If it is hot enough, it will pump out blackbody radiation in the form of light and even higher frequency electromagnetic waves. This was a problem for classical physics and its understanding of how waves worked. It made sense that the light should get brighter, but why did it change colour?

In theory, an ideal blackbody would absorb and emit radiation at all frequencies, but of course nothing in the real world is ideal. All the objects in the Universe are exchanging electromagnetic radiation with each other all the time. This is why nothing will ever cool to absolute zero, the theoretical lowest possible temperature at which a substance transmits no energy at all.

It's a catastrophe!

When the physicists of the late 19th century tried to explain their observations of blackbody radiation they ran into a problem. According to the laws of physics as they were understood at the time, a hot body ought to give off radiation at all frequencies, including short wavelength X-rays and gamma rays, as well as longer wavelengths like radio waves. Since there is effectively no upper limit on the higher frequencies, and therefore many more higher frequencies than there are lower ones, this would mean an infinite amount of waves being generated, all of them carrying energy. This came to be known as the ultraviolet catastrophe.

It seemed that there had to be a flaw somewhere in the current thinking about thermodynamics and electromagnetism, but what was it? No one could provide an answer to the problem. Then the German physicist Max Planck came along with a solution that was radical and stunning. It transformed physics.

How did Einstein learn about the quantum? | 55

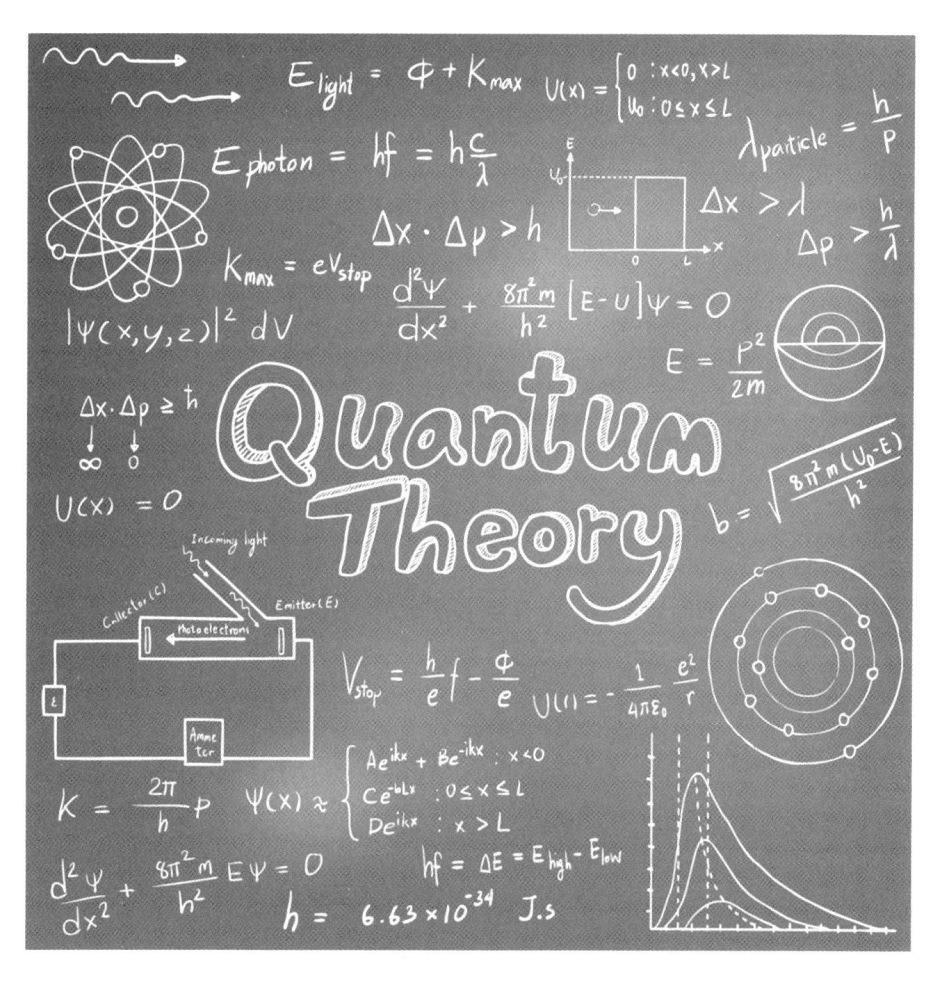

Into the quantum Universe

On 19 October 1900, Planck gave a speech to the German Physical Society that heralded the beginning of a new age for physics. He suggested that rather than being a continuous and infinitely variable quantity like a wave, energy in fact comes in packets. He called these packets quanta (singular quantum) from a Latin word meaning 'how much'. Energy could only be emitted or absorbed in whole quanta and each quantum had its own corresponding wavelength and frequency.

This explained why a blackbody would not give off radiation equally across the whole electromagnetic spectrum. It was much easier to emit an infrared quantum than an ultraviolet one, which required a lot more energy. The ultraviolet catastrophe was prevented from happening because, even though there are many more higher frequencies than lower ones, it required increasingly large amounts of energy to achieve them. A quantum of violet light, for example, has twice the frequency, and therefore twice the energy, of a quantum of red light.

Objects progress through the spectrum from infrared, to red, orange, blue and white hot as they heat up because the increase in temperature means that the particles that the object is made up of are becoming more energetic. This increase in energy is what allows higher frequency quanta to be formed. Planck suggested that the energy of a quantum was related to its frequency using the simple formula $E = \hbar v$ where E equals energy, v equals frequency and ℏ equals a value known as Planck's constant. The energy of a quantum of radiation can be calculated by multiplying its frequency by Planck's constant. Using his formula, Planck could predict accurately the total amount of energy in an oven at any given temperature.

There could be no doubt that Planck's solution worked; what the theory predicted agreed with what was found by experiment. Planck himself thought that his explanation was unlikely because it contradicted everything he had been taught and tried for years to disprove his own theory. But he accepted it as a convenient solution for what happened when matter absorbed or emitted energy, even though he was unable to give a good reason why it should be true. That reason would come just a few years later from Albert Einstein.

> **WHAT IS A CONSTANT?**
>
> In all the basic theories of physics, certain fundamental quantities appear that always remain the same whatever the conditions. These are called 'physical constants'. They include Planck's constant, h, linking energy and frequency in a quantum; c, the velocity of light, and G, the gravitational constant. Physical constants, it is believed, hold true under all circumstances and do not change.

When Max Planck was putting forward his theory of the quantum nature of energy 21-year-old Albert Einstein was giving private lessons as a maths tutor and working on what would become his first published paper, on the capillary effect. His reaction to Planck's theory was that, 'It was as if the ground had been pulled out from under us, with no firm foundation to be seen anywhere.' Planck was a big influence on Einstein's thinking. The two men spent a great deal of time together exchanging ideas over many years. When Planck died in 1947, Max Born remarked, 'It is difficult to imagine two men of more different attitudes to life … Yet what did all these differences matter in view of what they had in common – the fascinating interest in the secrets of nature.'

FIRING OUT PHOTONS

When atoms absorb sufficient energy the electrons surrounding the atomic nucleus can jump up into higher orbits. As the electron falls back to its original orbit it releases a quantum of electromagnetic energy, a photon. The amount of energy in the photon depends on how far back the electron falls. The further out from the nucleus the electron was boosted, the higher the energy of the photon that will be released when it returns to its original position. So, in blackbody radiation terms, the more you put in, the more you get out.

Chapter 5

What was Einstein's theory of the photoelectric effect?

Einstein took Planck's theory of the quantum and used it to explain the puzzling phenomenon of the photoelectric effect.

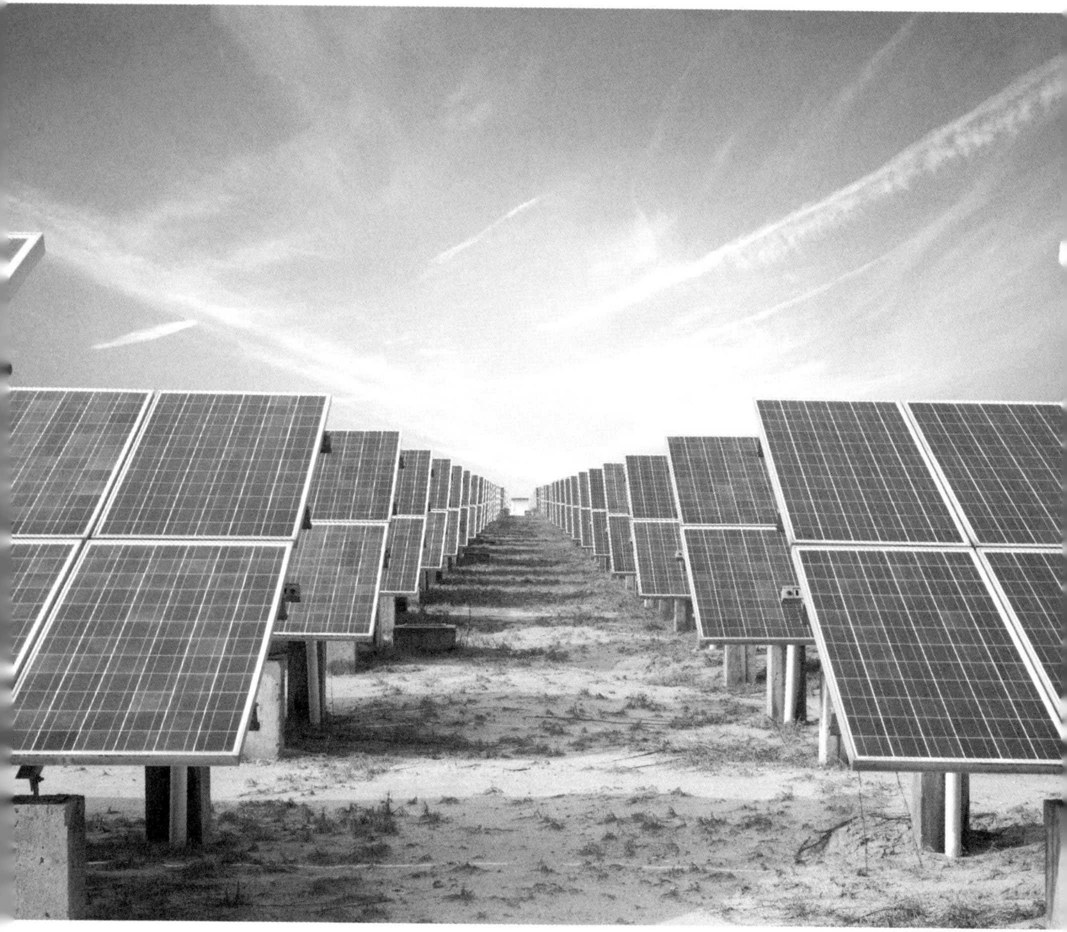

There was another curious aspect of radiation that needed an explanation. This was called the 'photoelectric effect'. It was known that if a beam of light was directed at certain types of metal electrons would be emitted. It's the science that underpins solar-generated electricity.

At first it seemed that the effect could be explained in terms of electromagnetism. The electric field part of the electromagnetic wave, it was assumed, gave the electrons the energy they needed to break free from the metal. But it soon became apparent that this couldn't be the whole story.

The energy of the electrons released depended on the frequency of the light – but not its intensity. It seemed that a brighter light should have more energy and so produce higher energy electrons but, no matter how bright the light used, the electrons that emerged still had the same energy. It was only by shifting the light up in frequency, from red to violet and then ultraviolet, that higher energy electrons were emitted. A brighter light produced more electrons but their energy stayed the same. If the light was of a low enough frequency no electrons would be emitted at all, even if the light was blindingly bright. The wave theory of light failed to explain these findings.

Light quanta

Einstein was intrigued by the phenomenon. In 1904, he wrote to a friend that he had 'found in a most simple way the relation between the size of elementary quanta of matter and the wavelengths of radiation'. In a paper he published in the journal *Annalen der Physik* (*Annals of Physics*) in March 1905, the first in that remarkable year, Einstein took the findings of the experiments on the photoelectric effect and married it to Planck's theory and came up with a result that would win him the 1921 Nobel Prize.

Einstein began by looking at the differences between particle theories, including those that described the behaviour of gases, and wave theories, and those describing the behaviour of electromagnetic radiation. He emphasized that, though he was about to put forward a strong case for light to be considered as a stream of particles this did not mean that the wave theory would have to be abandoned – it had worked well and would continue to be useful, he said.

Describing light as packets of energy, Einstein wrote that: 'When a light is propagated from a point, the energy is not continuously distributed over an increasing space but consists of a finite number of energy quanta which are localized at points in space and which can be produced and absorbed only as complete units.' Einstein's biographer Walter Isaacson described this as 'perhaps the most revolutionary sentence that Einstein ever wrote'.

Einstein's approach was simple yet compelling. He compared the formulae that described how the particles in a gas behaved as it changed volume with those that described similar changes as radiation spread through space.

He found that both obeyed the same rules. The maths underpinning the behaviour of the gas was the same as that for the radiation. It also gave Einstein a way to calculate what the energy of a light quantum of a particular frequency would be. His results agreed with Planck's findings.

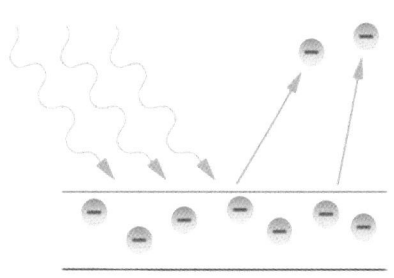

The photoelectric effect explained

Next, Einstein showed how the existence of light quanta explained the photoelectric effect. As Planck had shown, the energy of a quantum was determined by multiplying its frequency by Planck's constant. If a single quantum transferred all of its energy to a single electron then the higher the energy the quantum possessed the higher the energy the emitted electron would have. Increasing the intensity of the light would produce more electrons but it wouldn't increase their energy. This was in line with the observations of the photoelectric effect made in the laboratory.

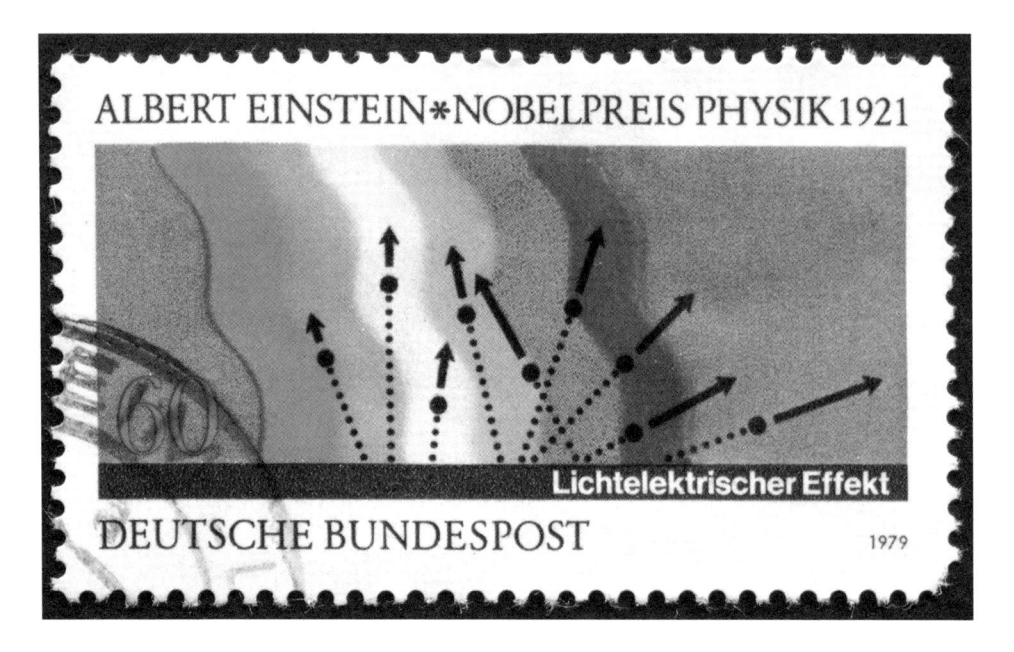

Quantum reality

As far as Planck had been concerned, the quantum was little more than a mathematical contrivance that he used to make the equations work out. But to Einstein it was a physical reality, a feature of the Universe as it really was. By 1916, experiments had confirmed that Einstein was right and scientists were having to rethink their ideas on the nature of light.

QUANTUM STRANGENESS

Thomas Young demonstrated that light was a wave by showing how light passing through two close-together slits formed interference patterns. But supposing light is a stream of particles, what would happen if we sent just one quantum at a time towards the slits? You'd imagine that the interference patterns would be replaced by two bright patches in line with the slits. But spookily, even if we only fire off one photon a second, the interference patterns still form. Pause to think about that. Even if subsequent photons are fired off after the earlier ones have hit the screen they still somehow 'know' where to go to build up the interference pattern! How does that work? As the great physicist Richard Feynman said '... a lot of people understood the theory of relativity in one way or another ... On the other hand, I can safely say that nobody understands quantum mechanics.'

For the next 20 years, Einstein struggled without success to solve the dual nature paradox of light. Even towards the end of his life, in 1951, Einstein wrote in a letter to his friend Michele Besso that 50 years of pondering had brought him no nearer to finding an answer to the question 'What are light quanta?' 'Nowadays every Tom, Dick and Harry thinks he knows it,' Einstein wrote, 'but he is mistaken ... No one really knows exactly what light is!'

Wave–particle duality

What became apparent was that, bizarre and headstretching as it might be, the wave theory and the particle theory of light are both correct. Whether light is a wave or a particle seems to depend on how you look at. All we can do is describe how light behaves under different conditions – sometimes it behaves like a wave and sometimes it behaves like a stream of particles. Sometimes it seems to be both at the same time. We have no single model that can describe light in all its aspects. It's easy enough to say that light has 'wave–particle duality' and leave it at that, but what that actually means is something that no one can answer satisfactorily.

LIGHT IS A particle!

Chapter 6
How did Einstein prove the existence of atoms and molecules?

At the beginning of the 20th century scientists were still arguing about whether atoms actually existed – Einstein took the work of a Scottish botanist and showed that they did.

In May of 1905, the *Annalen der Physik* received another paper from Einstein. The subject this time – the kinetic theory of gases – was grounded in classical physics, but Einstein would use it to establish for the first time that atoms and molecules were a physical reality.

Which way are we going?

Earlier we looked at Newton's laws of motion, which allows us to calculate the path of a comet-catching space probe, or the flight of a cricket ball into the wicketkeeper's gloves. Using Newton's laws, if you know how an object is moving now you can work out how it was moving in the past, and how it will move in the future.

The funny thing about the laws of motion is that they aren't time dependent. If you look at a film of a comet flying through space there is no way that Newton's laws can tell you what direction the film is running in. Newton's laws are time reversible, they work just as well in either direction, whether you are calculating motion forwards into the future or backwards into the past.

Of course, for most situations common sense and experience tell us whether an event is taking place forwards or backwards in time. If I showed you a film of a broken egg putting itself back together and rising up off the floor and into my hand you'd know right away I was running it in reverse. Newton's laws don't say that this is impossible, but it's fairly certain you'll never see it happen because it is very, very, very unlikely.

The reversibility paradox

According to the kinetic theory of gases, heat is a measure of the motion of atoms. The more agitated the atoms, the higher the heat. Ludwig Boltzmann used the kinetic theory to resolve the so-called 'reversibility paradox' in physics. This arose from the second law

reversible

irreversible

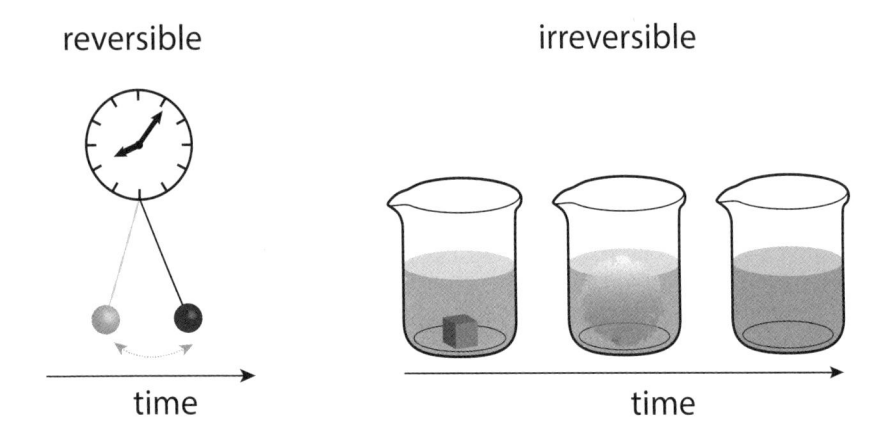

time · time

of thermodynamics, which states that physical systems tend to become more disordered and dictates that most natural processes are irreversible. The Universe, it seems, proceeds inexorably from a state of low entropy (order) to a state of high entropy (disorder), which appears to contradict the time reversible nature of Newtonian mechanics. The idea of an 'arrow of time', pointing from the past to the future, was first introduced by the astronomer Sir Arthur Eddington. He played an important part in Einstein's story and we'll return to him later.

Boltzmann solved the paradox by determining that the second law was about probabilities. All of the countless atoms and molecules that make up a broken egg or any other object are in constant random motion. There is a vanishingly small, but not impossible, chance that the molecules will all move in just the right direction to reassemble the egg. But this event is just so utterly improbable that the breaking of the egg is effectively irreversible.

Albert Einstein read Boltzmann's theory of gases – portraying a gas as a collection of countless molecules bouncing around randomly – and declared it 'absolutely magnificent'. Between 1902 and 1904, Einstein was also working on the second law of thermodynamics

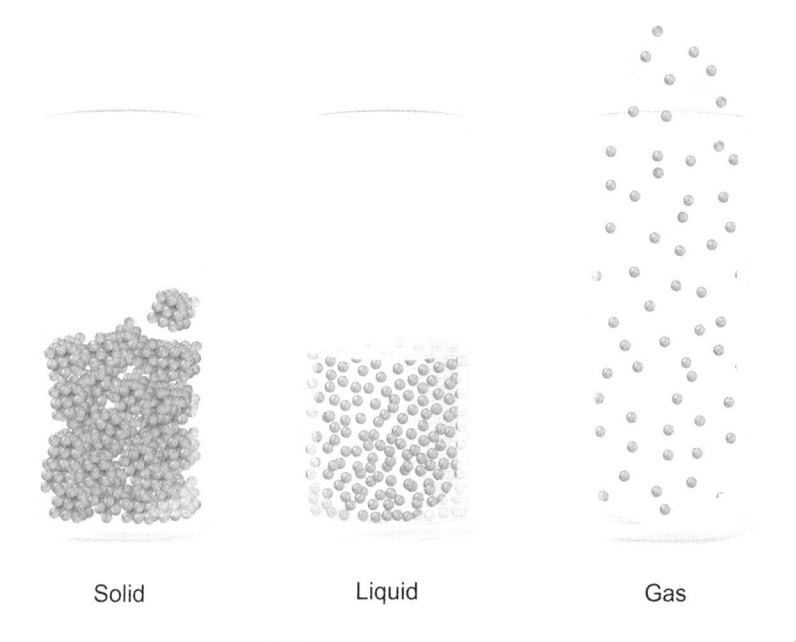

Solid   Liquid   Gas

$N_A =$ 6.02 x 10^{23}

and developing a 'general molecular theory of heat' using statistics and mechanics that would 'close the gap', as he put it, and extend Boltzmann's work on gases to other materials. He presented his statistical molecular theory for his doctoral dissertation at the University of Zurich. In the dissertation, Einstein described a new theoretical method for determining the size of molecules and for calculating Avogadro's number, which is the number of atoms or molecules in a set amount of a substance.

In a separate paper, published in May 1905, he applied the molecular theory of heat to liquids and ended up solving the puzzle of 'Brownian motion'.

Brownian motion

In 1827, the Scottish botanist Robert Brown noticed that flower pollen suspended in water moved about randomly, seemingly being

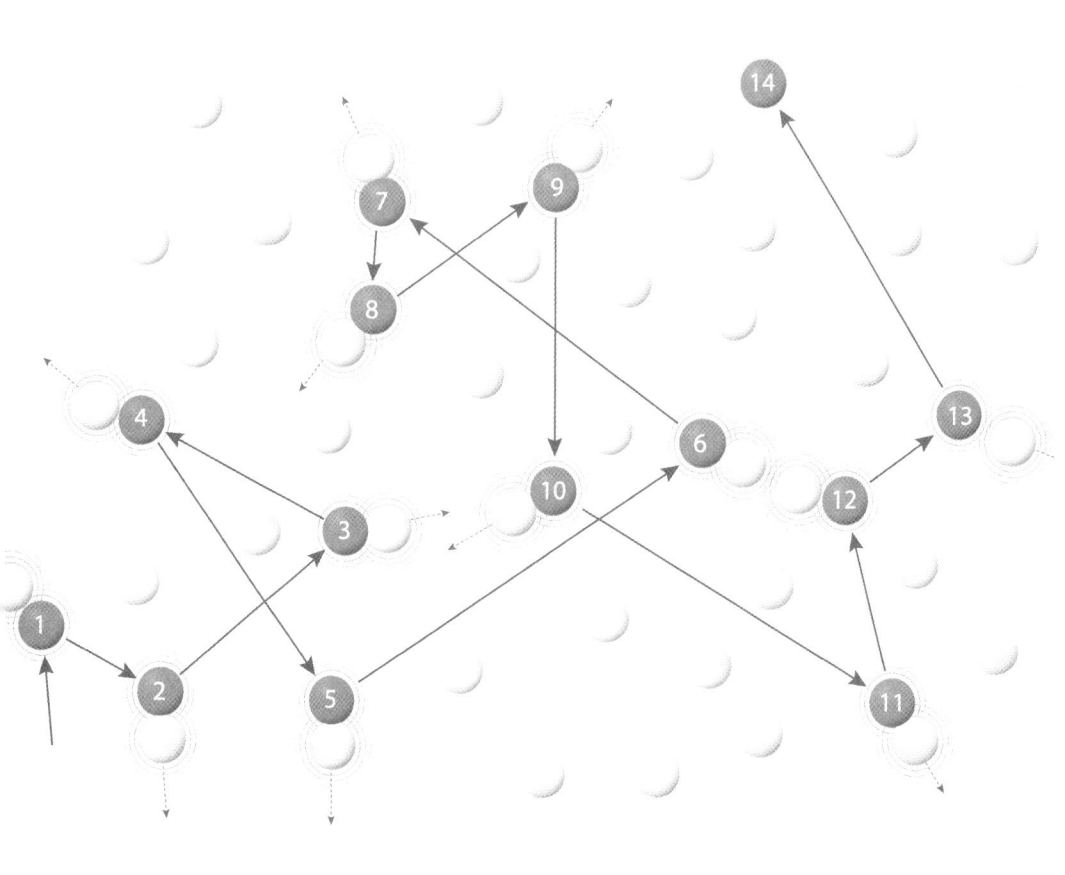

How did Einstein prove the existence of atoms and molecules? | **73**

nudged by unseen forces. Other researchers had noticed this odd phenomenon but Brown was the first one to study it. To begin with, Brown thought it had something to do with the pollen being alive, but experiments showed him that it wasn't just pollen grains that moved around in this way – any particles of a similar size, be it chips of granite or particles of smoke, suspended in liquid showed just the same odd motion, which came to be named after Brown.

Einstein didn't specifically set out to explain Brownian motion. He didn't even mention Brownian motion in the title of his paper, calling it 'On the Motion Required by the Molecular Kinetic Theory of Heat of Particles Suspended in Fluids at Rest'. In his paper, he wrote: 'It is possible that the motions to be discussed here are identical with so-called Brownian molecular motion; however the data available to me on the latter are so imprecise that I could not form a judgement on the question.'

Einstein wanted to find evidence for the existence of atoms and molecules. He wanted to show how the actions of molecules could be visibly demonstrated; explaining Brownian motion was just incidental to that; he didn't even realize how well known a phenomenon Brownian motion was. Einstein reasoned that if the molecules in a liquid were moving randomly, just as the molecules in a gas did, then every so often they would jostle against the tiny pollen grains in sufficient numbers to make them move. Einstein explained the motion in detail, using theoretical knowledge and the data from experiments, coupled with the powerful statistical tools he had mastered in his earlier dissertation, to accurately predict how far the particles would travel in the course of their irregular, random motions.

When Einstein's paper on Brownian motion first appeared in 1905, scientists were still debating the very existence of atoms and molecules. Some scientists such as the physicist Ernst Mach (who gave his name to the speed of sound) and the physical chemist Wilhelm Ostwald were among those who chose to argue against the atom. They took the view that thermodynamics was to do with the way energy changes from one form to another and there was no need to explain it in terms of randomly moving invisible atoms.

Mach had an influence on Einstein's thinking in another way. He declared that it was impossible to define Newton's concepts of absolute time and space, calling it a 'conceptual monstrosity'. Einstein would overturn these ideas altogether.

Within months of the publication of Einstein's paper, his predictions had been confirmed by experiment. The French physicist Jean-Baptiste Perrin used the newly invented ultramicroscope to verify Einstein's ideas and was awarded the 1926 Nobel Prize for physics for doing so. The evidence Perrin found to support Einstein's theory was so compelling that the reality of atoms and molecules had to be accepted. Physicist Max Born wrote: 'I think that these investigations of Einstein's have done more than any other work to convince physicists of the reality of atoms and molecules.'

It is a mark of Einstein's genius that, at the same time as he was proving the existence of atoms, he was also working out the consequences of travelling at the speed of light. Just a few days after the publication of the paper on molecular motion, he was telling a friend that he was about to modify 'the theory of space and time'.

Chapter 7

What was Einstein's theory of special relativity?

Einstein showed that, because the speed of light didn't change, everything else had to.

The idea of relativity in physics is a fairly straightforward one. What it says is that the laws of physics apply and remain the same for all freely moving observers, whatever their speed of motion. But what do we actually mean by being in motion?

Einstein's theory of special relativity, developed in 1905, was special in the sense that it concerned the special state of objects in uniform motion, moving at a constant velocity and direction relative to one another. Physicists call this an inertial frame of reference. As Newton pointed out in his first law of motion, an inertial state is the default for any object not being acted on by a force. Inertial motion is simply motion at a uniform speed in a straight line. It would be another ten years before Einstein formulated his general theory, which also encompassed objects in accelerated motion.

In order to measure anything, be it time, or distance, or mass, you have to have something to measure it against. An object is only faster, bigger or heavier relative to something else. If you have nothing to compare it with it's pretty meaningless to say 'the big, heavy whatchamacallit was moving really fast!' One of the reasons we developed a system of weights and measures was to have a way of comparing similar objects and quantities so we could all agree that one thing was bigger or heavier than another. None of our units of measurement are absolute, they all have to be defined by reference to something else.

Galileo

Back in 1632, Galileo explored the idea that all motion is relative and that it only makes sense to speak of something moving in relation to something else. In his *Dialogo sopra i due massimi sistemi del mondo* (*Dialogue Concerning the Two Chief World Systems*), Galileo's intent was to defend the idea that the Earth does not sit motionless at the centre of the Universe. If the Earth was moving round the Sun, as Copernicus had suggested, then surely, said the critics, we'd feel it as we rushed through space.

Galileo tackled the argument by imagining the situation of a person inside a windowless cabin on a ship sailing on a perfectly smooth lake at a constant speed. He asked, is there any way in which the passenger can determine that the ship is moving without going on deck?

If the ship continues to move at constant speed and direction the passenger will not feel its motion. It's just the same for a modern-day passenger travelling on a smoothly running train or an aircraft cruising through the sky. Without looking out of the window to watch the world going by it's impossible to tell you're going anywhere.

Galileo wondered if there were any experiment that could be carried out on the ship that would give a different result to the same experiment being carried out on the shore and therefore give some indication that the ship was in motion. Galileo concluded that it was impossible. Any mechanical experiment performed inside the ship, provided it was moving at a constant speed in a constant direction, would give exactly the same results as a similar experiment carried out on shore. From these observations Galileo put forward his relativity hypothesis:

Any two observers moving at constant speed and direction with respect to one another will obtain the same results for all mechanical experiments.

Frames of reference

An important consequence of this is that velocity can only be measured with reference to something else and that the measurement we get changes if we measure the velocity from a different reference point. To say something is moving only has meaning if you can say what it is moving relative to. If two people sit opposite each other on a train and one tosses the other an orange, it travels through the air at a few kilometres per hour, but someone standing at the side of the tracks would see orange, train and passengers rush by at a hundred kilometres per hour! The speed at which you perceive an object to be moving depends on how fast you yourself are moving relative to that object.

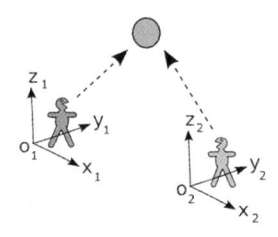

The idea that motion has no meaning without a frame of reference is fundamental to Einstein's theories of relativity. Before Einstein, it was believed that there was such a thing as absolute motion, meaning that an object could be said to be moving without reference to anything else. This idea required that there must also be a state of absolute rest in space (either something is moving or it isn't). These ideas were formulated by Newton, who wrote: 'Absolute motion is the translation of a body from one absolute place into another; and relative motion, the translation from one relative place into another.' Einstein's theory of special relativity did away with this idea of absolute rest and absolute motion. On one, perhaps apocryphal, occasion, Einstein supposedly asked a bemused ticket inspector, 'Does Oxford stop at this train?'

Introducing special relativity

Einstein's third paper of 1905 was called 'On the Electrodynamics of Moving Bodies'. It starts with a very simple example, well known to Michael Faraday and the other Victorian researchers into electromagnetism: an electric current is generated if a magnet is moved inside a coil of wire, and the same current is produced if the magnet remains fixed and the coil is moved. Einstein was very familiar with electricity – he often helped his engineer uncle Jakob, who had also been responsible for introducing the young Einstein to the delights of algebra, to tinker with the coils and magnets in a generator. Einstein's job in the patent office also meant he was regularly examining a variety of electromechanical devices.

Since Faraday's time, it had been assumed that there were two different explanations at work – one for the moving magnet producing a current and another for the moving coil producing the current. But Einstein wasn't having any of that and said that it didn't matter which was moving, it was their movement relative to each other that generated the current. As he said, 'The idea that these two cases should essentially be different was unbearable to me.'

The distinction between moving magnet and moving coil depended on the view still held by most scientists that there was a state of absolute rest with respect to the ether, that mythical and mysterious substance that Michelson and Morley had failed to find any evidence for.

The example of the magnet and coil, along with every observation that had been made about the nature of light, led Einstein to conclude that the idea of absolute rest was flawed and unnecessary.

In a casual sentence, he dismissed the idea of the ether entirely: 'The introduction of the "light ether" will prove to be superfluous… the view developed here will not require a "space at absolute rest".' He set out his 'Principle of Relativity':

'The same laws of electrodynamics and optics will be valid for all frames of reference for which the laws of mechanics hold good.'

Another way of saying this is that the laws of physics are the same in all inertial frames of reference. No matter whether you're travelling fast or slow, this way, that way, forwards or backwards, the laws remain the same, which means that any experiment carried out will produce results that are in accordance with the laws. And that's exactly what Galileo was saying back in 1632 – Einstein and Galileo are both in agreement that no experiment can determine the motion of the observer in an inertial frame.

It's important to keep in mind that special relativity only applies to objects moving in an inertial frame. Once the object changes direction or speeds up or slows down then it can be determined to be in motion. We feel it when a car accelerates or a plane begins its descent. There is no need to say an object is accelerating with respect to anything else.

The constancy of light

Once Einstein had adopted his Principle of Relativity, he realized that it was impossible for both Newton and Maxwell to be right. Einstein decided to side with Maxwell and began his challenge of 200 years of Newtonian physics.

In 1940, Einstein wrote:

'The precise formulation of the time-space laws was the work of Maxwell. Imagine his feelings when the differential equations he had formulated proved to him that electromagnetic fields spread in the form of polarized waves, and at the speed of light! To few men in the world has such an experience been vouchsafed ... it took physicists some decades to grasp the full significance of Maxwell's discovery, so bold was the leap that his genius forced upon the conceptions of his fellow workers.'

Einstein asked the question: Does light behave the same way as everything else? Is the speed of light also dependent on the motion of the observer? This brought Einstein to the second postulate, called the light postulate, that he founded his theory on: that the speed of light is a constant. Some things may be relative, but the speed of light is absolute.

Light, said Einstein, always travelled at a constant velocity that was independent of the velocity of the object emitting the light. This made little sense in Newtonian terms, in which speeds were added together. For example, a fast bowler can bowl a cricket ball faster by adding the speed of his run in to the speed at which the ball is released from his hand. But a beam of light projected from a fast-moving aircraft would still travel at the same speed as one projected from a (relatively) stationary mountaintop below it.

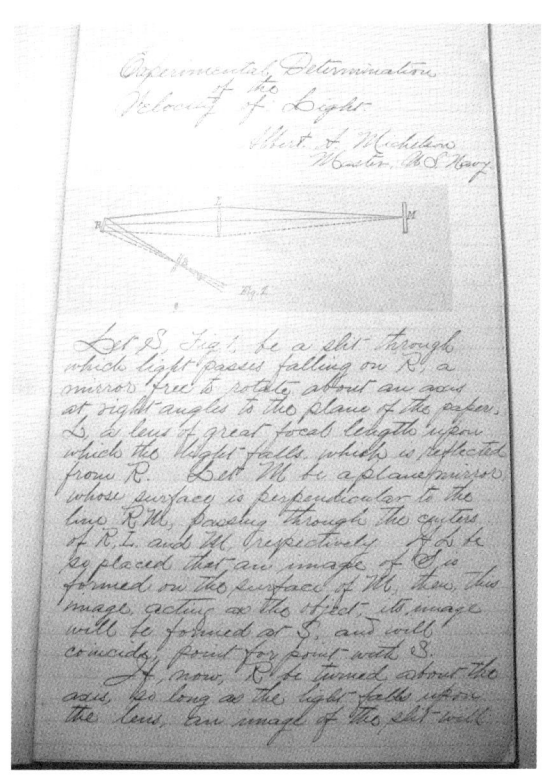

This was what Michelson and Morley had found when they discovered that the speed of light was always the same no matter how they measured it. Oddly, Einstein makes no mention of Michelson and Morley. He even claimed at one point that in 1905 he hadn't even heard of their experiment, though in later years he often contradicted himself. For instance, on 15 January 1931, at a dinner given at the California Institute of Technology, Einstein publicly addressed Michelson for the first and – what turned out to be – the last time, since Michelson died a few months later: 'I have come among men who for many years have been true comrades with me in my labours. You, my honoured Dr Michelson, began with this work when I was only a little youngster, hardly three feet high. It was you who led the physicists into new paths, and through your marvellous experimental work paved the way for the development of the theory of relativity.'

The idea that nothing in the Universe can travel faster than light is central to Einstein's special theory of relativity. But why does light always travel through a vacuum at nearly 300,000 kilometres per second (km/s)? Why not faster or slower?

Simply put, it's because that's the answer we get when we solve Maxwell's equations. In Maxwell's equations, the speed of

> **LIGHT SPEED DEFINED**
>
> In 1983, the General Conference on Weights and Measures officially defined the speed of light to be:
>
> $c = 299{,}792{,}458$ m/s.
>
> Scientists use 'c' as the symbol for the speed of light from the Latin word 'celeritas', which means 'swiftness'.
>
> At the same time the metre came to be defined as the distance light travels in one 299,792,458th of a second.

electromagnetic waves is a constant defined by the properties of the vacuum of space through which the waves move. It is not measured relative to anything else as any other speed would be. The nature of the Universe and the behaviour of electric and magnetic fields dictate what the speed of light must be.

Because Maxwell's equations, which determine the speed of light, hold true in any inertial frame, two observers moving relative to each other, each measuring the speed of a beam of light relative to themselves, will both get the same answer – even if one is moving in the same direction as the beam of light and one away from it. Everything else in the special theory of relativity derives from that one simple fact. The constancy of the speed of light produces many seeming paradoxes that turn our notion of space and time on its head, and we'll be looking at some of those next.

> **SLOW LIGHT**
>
> The 'speed of light' is generally taken to refer to the speed of light in a vacuum. Light doesn't always travel this fast though. It slows down as it passes through transparent media such as air, water or glass. The speed of light through water is about 75 per cent of its speed in a vacuum but that's still about 225,000 km/s so you won't notice the difference. The ratio by which it is slowed down is called the refractive index, discovered by Jean Foucault in 1850.

Chapter 8
What were Einstein's ideas about time?

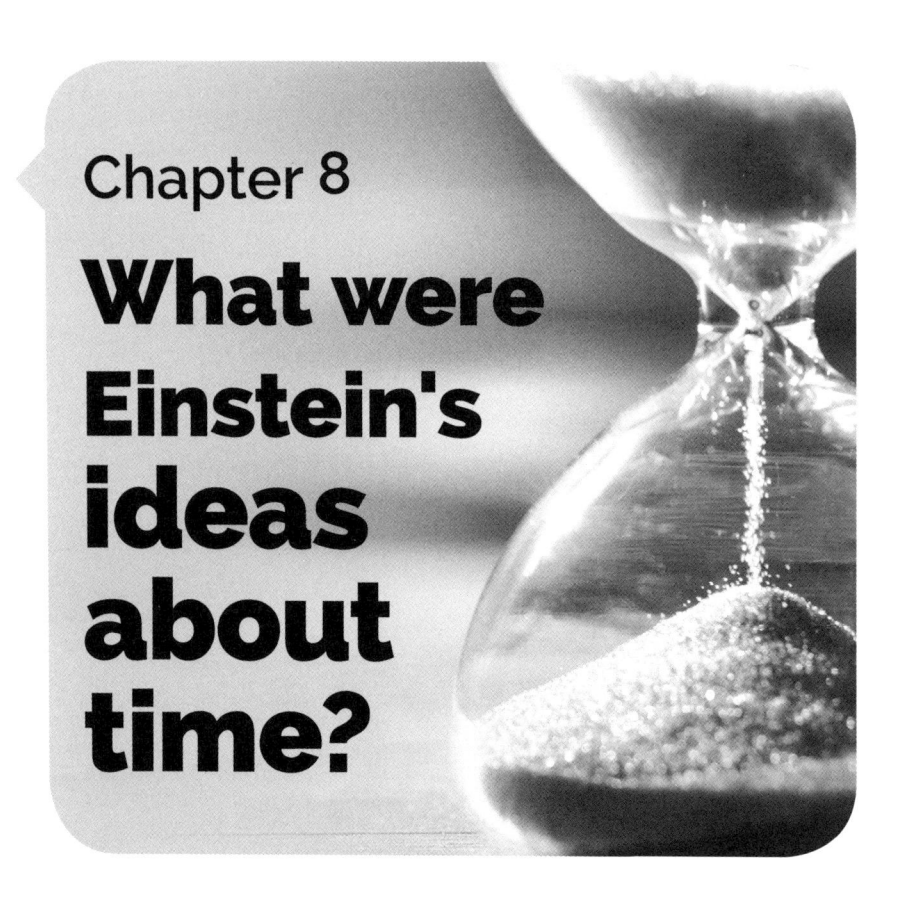

The consequences of special relativity turned our notion of time on its head.

In a talk he gave in 1922, Einstein related some of the difficulties he had with explaining why the speed of light was the same to all observers. 'My solution was really for the very concept of time, that is, that time is not absolutely defined but there is an inseparable connection between time and the signal velocity [of light]. With this conception, the foregoing extraordinary difficulty could be thoroughly solved. Five weeks after my recognition of this, the present theory of special relativity was completed.'

Imagine you're in a space station and you fire off a laser signalling device at a spacecraft that's heading away from you at half the speed of light, around 150,000 km/s. Common sense would tell you that the laser light should reach the spacecraft at half-light speed because it has to catch up with the spacecraft, but common sense is wrong. The beam will still arrive at the spacecraft at approximately 300,000 km/s. If you recall your physics lesson, velocity equals distance travelled divided by time taken to get there, or as an equation:

v = d/t

It is, in other words, a measure of space divided by time. So it seems that in order for the speed of light always to remain the same for all observers (that is, that we always get the same answer for v), we have to do some tinkering with d and t. If light speed is unalterable, time and space must change.

Absolute time

'Time exists in and of itself and flows equably without reference to anything external.' So wrote Isaac Newton. In Newton's world time always ticked by at the same pace wherever you measured it. If our timepieces are both accurate, then five seconds for me will be five seconds for you. Except they won't. Go back to the space station for a moment and the laser signalling device. For you on the space station and the pilot in the spacecraft to agree on the speed the beam reaches the spacecraft you have to agree on the time it takes to get there. Since the speed of light is always the same for the spacecraft pilot, to get the same answer the spaceship's clocks have to be running slower.

The twins paradox

Imagine that the pilot of the swiftly departing spacecraft, now accelerating to 0.99 per cent of light speed, is your twin setting out on a voyage of exploration through space that will take one year of ship's time to complete. When your twin returns, he or she will be, as you'd expect, a year older. But how much older will you be?

As you track the ship's progress you will notice something rather odd occur. The clocks onboard the ship are ticking more slowly than the clocks aboard the space station (or perhaps it's that yours are running faster, relatively speaking). The upshot of this is that, by the time your twin's ship returns to the space station, seven years have gone by for you. (The actual time will depend on just how close to light speed the spaceship travels – at half light speed an hour on the ship would be equivalent to 69 minutes on the space station; very close to light speed the difference might be measured in thousands, or even millions, of years.) You now have a twin who is six years youngwr than you are. But that isn't the paradox.

Recall that Einstein said that all motion is relative. In his first example, it didn't matter whether the coil or the magnet moved in order to produce a current; either way, they were moving relative to each other. By the same token, your twin aboard the spaceship can look back at the space station as it falls far behind and say that it's you who is retreating at close to light speed. That being the case, your twin will see your clocks moving more slowly! So, which of you is ageing more slowly? Both of you? Or neither?

You might think that it all balances out and you'll still be the same age when you're reunited. It's a fair assumption. But it's wrong. The spaceship pilot twin really will have aged less. Physicist Herbert Dingle, writing in the 1960s, argued that the twin paradox revealed an inconsistency in special relativity. Today, most scientists are in agreement that relativity theory can solve the puzzle but there is disagreement as to whether the solution can be found in special relativity or only in general relativity. Einstein himself said the solution to the paradox requires general relativity. That being the case, we will return to the solution later.

Faster, and slower

According to special relativity, the faster you travel through space, the slower you travel through time. As you approach the speed of light the intervals between events lengthen, so time seems to slow down. This phenomenon is called time dilation and it actually happens, it's not just a question of different observers having a different perception. If an object could achieve the speed of light then time would appear to stop entirely. In experiments, such as those carried out in the Large Hadron Collider at CERN, where atomic particles are smashed into each other at significant fractions of the speed of light, the effects of time dilation must be taken into consideration if the results are to make any sense.

THE NATURE OF TIME

Time passes. Time can be long or short; time can hang heavy or fly by; we take time and we try to make time; we save time and we wonder where the time has gone. Nobody really knows what time is. In 1905, the French physicist Henri Poincaré argued that time is something we've invented for our convenience rather than a feature of reality. He declared that there were no tests that would tell us anything about the nature of time, and that we should just adopt whatever concept of time makes for the simplest laws of physics.

We might think of time as the thing that separates one event from another, that can tell us how long an event lasted, and which event came first. For example, it takes a top sprinter ten seconds to go from the starting blocks to the finishing line in a 100-metre race. We can measure the time taken in fractions of a second so we know who won and whether or not it was a record time. But what is a second? We can define it in terms of the vibrations of an atom but each vibration is just another event in time. We're no closer to understanding what time is. If nothing happened anywhere, if there were no events occurring at all, would time still 'flow equably' regardless? Perhaps time is just things happening. And, as Einstein showed, things can happen at different rates.

What were Einstein's ideas about time?

The end of simultaneity

There was one effect of special relativity that Einstein thought particularly important. This is the relativity of simultaneity. Two events that appeared to one observer to have occurred simultaneously might not appear to have done so to a second observer who was moving relative to the first. In addition, according to Einstein, there is no way in which we can say one observer is correct and the other one wrong. They are, in fact, both right!

Einstein explained the conundrum in terms of a thought experiment. Imagine that you are watching a thunderstorm and suddenly two buildings that you know to be equidistant from you are struck by lightning. You would say that the two had been struck simultaneously. Now suppose a bus goes by. If the lightning strikes occur while a passenger on the bus is level with you, he will doubtless agree that the two strikes occurred simultaneously. Now imagine that the bus is moving towards one building and away from the other. In that case the light from the strike on the second building will take longer to reach the passenger than the light from the one he is moving towards. He will not then see the two strikes as simultaneous.

As we have seen, the principle of relativity says that there is no way to insist that you are at rest and the passenger on the bus is in motion. You are simply in motion relative to each other. There is, therefore, no 'right' answer to whether or not the strikes occurred simultaneously.

The end of simultaneity is another nail in the coffin of absolute time. Two observers in relative motion will have clocks that tick at different rates; the effect becomes more marked approaching the speed of light but it is there, if infinitesimally, at low relative speeds too. Time passes differently for all moving reference frames. As physicist Werner Heisenberg put it: 'This was a change in the very foundation of physics, an unexpected and very radical change that required all the courage of a young and revolutionary genius.'

Chapter 9

How did Einstein explain the Lorentz-Fitzgerald contraction?

It wasn't enough for Einstein to bend time to fit his theory, he shrank space as well.

Another of the strange consequences of the speed of light remaining constant to all observers is that a moving object appears to shrink along the direction of motion. At the speed of light, the length of the object would be zero. This phenomenon is called the Lorentz–FitzGerald contraction after the two physicists who proposed it in 1889 as a solution to the failure of the Michelson–Morley experiment. It was Einstein who showed the phenomenon was real but a consequence of the properties of space and time and not an actual physical compression.

Just follow the bouncing beam

Because time slows down the faster we go it follows from this that we have to physically contract too. To see how this works, imagine that a spacecraft has a mirror mounted on each end and that a pulse of light bounces between the two mirrors. What happens to the bouncing beam as the spacecraft approaches the speed of light?

For a 150-metre-long ship at rest, the return journey for the light beam will take roughly a millionth of a second. However, at 99.5 per cent of the speed of light, time is slowed by around a factor of 10, which would mean the round trip journey time, as measured by an observer, is now a hundred thousandth of a second. The trouble is, the pulse heading from the back to the front will take longer to get there, because the mirror is retreating from it at close to the speed of

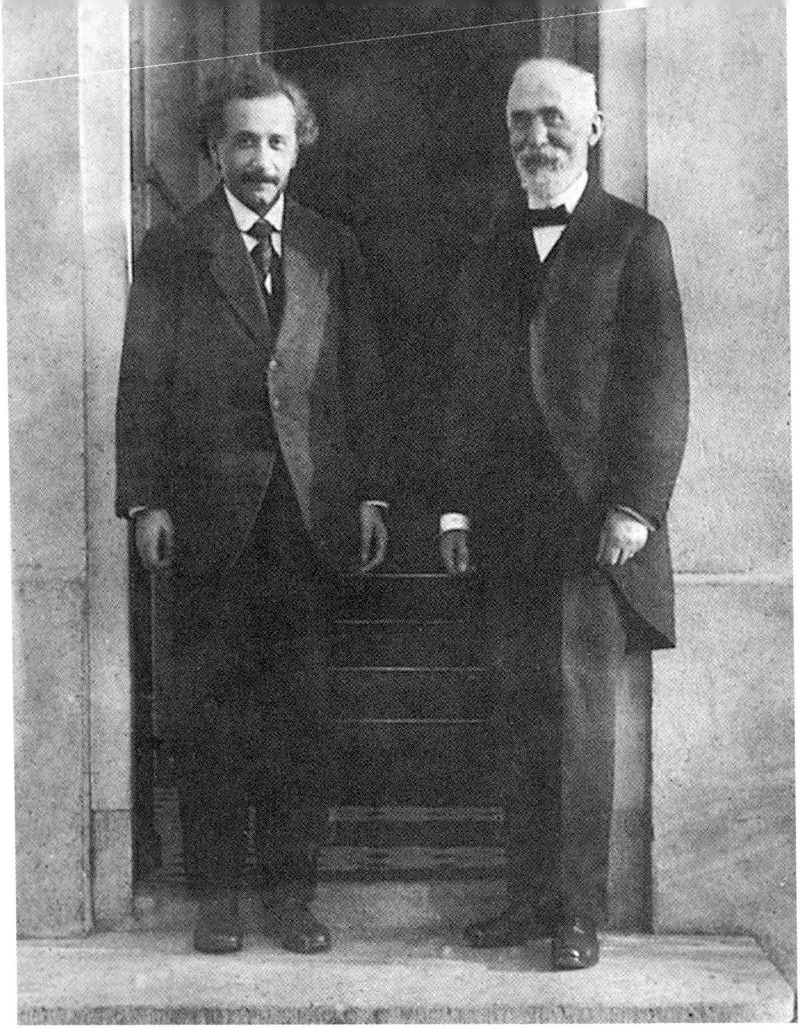

light. The return trip will take a much shorter time because the rear mirror is rushing towards the light beam. But no matter whether it's a retreating mirror or an advancing mirror, the light beam will always reach it at the same speed, roughly 300,000 km/s, because the speed of light, remember, doesn't change.

Einstein asked himself a question. If I could fly at the speed of light and held a mirror in front of myself, would I see my reflection? How would the light reach the mirror if it was retreating at light speed? It was thought experiments like this that built the foundations of special relativity. The answer is that he would see his reflection because no matter how close he is to light speed the light travelling from him to the mirror and back again will always be travelling at the same 300,000 km/s.

In order to balance the books, as it were, and ensure that light speed is always measured to be the same, not only does time have to slow down, the distance travelled by the light beam has to decrease also. At roughly 99.5 per cent of light speed, the distance is reduced by a factor of 10 – the same proportion as the time dilation effect.

The spaceship and its crew don't shrink in size. The object in motion is only shortened in the direction of its motion; those dimensions that are perpendicular to its motion remain the same. The result is that, to an observer at rest relative to the moving object, it becomes distorted from its resting shape.

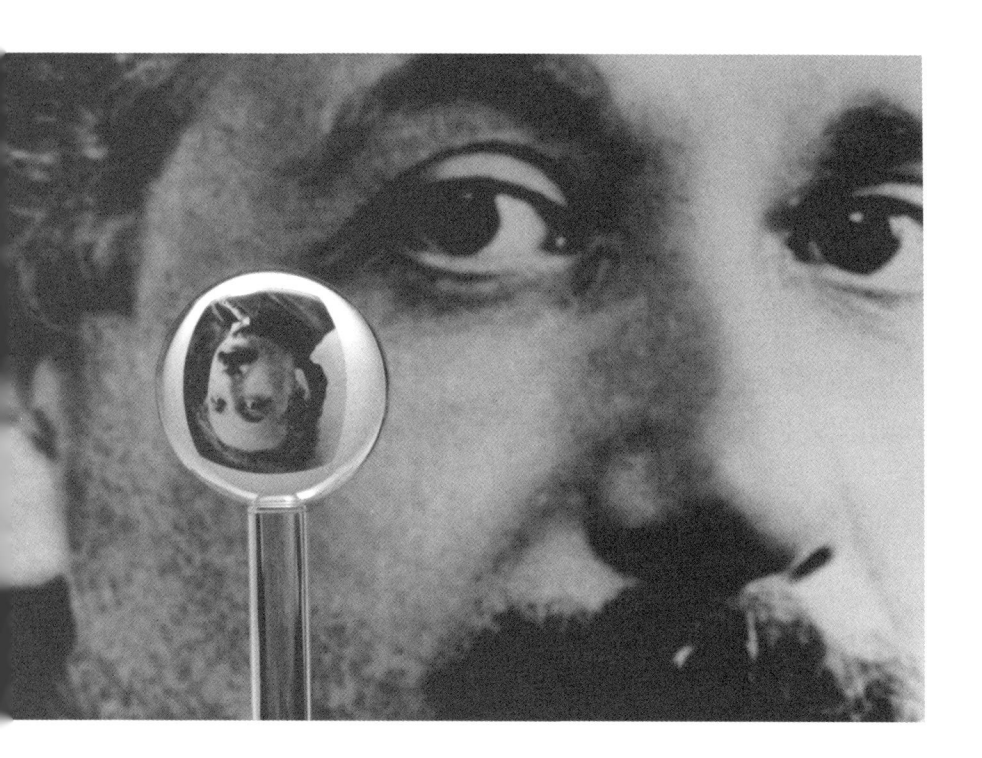

The change in length won't be apparent to the crew of the ship. The distortion will only be apparent to an observer who is relatively at rest compared to the ship. From the perspective of the ship's crew, it is you, the observer, who will appear to have contracted as, relatively speaking, it is you who is streaking past them.

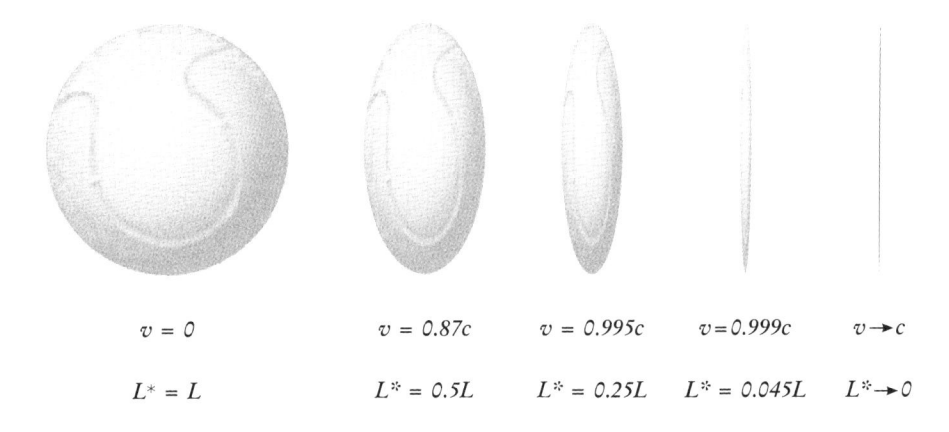

$v = 0$	$v = 0.87c$	$v = 0.995c$	$v = 0.999c$	$v \to c$
$L^* = L$	$L^* = 0.5L$	$L^* = 0.25L$	$L^* = 0.045L$	$L^* \to 0$

To boldly go...

As the late Douglas Adams so memorably put it, 'Space is big. Really big. You just won't believe how vastly hugely mindbogglingly big it is.' Even light, the fastest thing in the Universe, takes over four years to traverse the distance between the Sun and the next closest star. Yet should it ever become technically possible to attain near-light speed, travel to the stars for these swiftly moving pioneers would take a lot less time. How is that possible?

It's another consequence of the distance to be travelled shrinking at high speeds. Imagine that the spaceship moving at light speed travelled along a rail that stretched from star to star. The faster the ship travelled, the shorter the rail would appear to become and, therefore, the shorter the distance they would have to cover to reach their stellar destination. At 99.5 per cent of light speed, the journey to the nearest star would take about five months rather than four years. The closer the ship got to the speed of light, the shorter the journey time would be.

There is a downside to this, of course. The ship's clocks, you will remember, are ticking ten times more slowly than the relatively motionless (nothing is actually motionless, of course) clocks of their friends and relatives back on Earth. So, although only five months will pass according to ship's time, by Earth time it will still take four years to complete the journey. The faster the ship travelled the more extreme the discrepancy between ship time and Earth time would become.

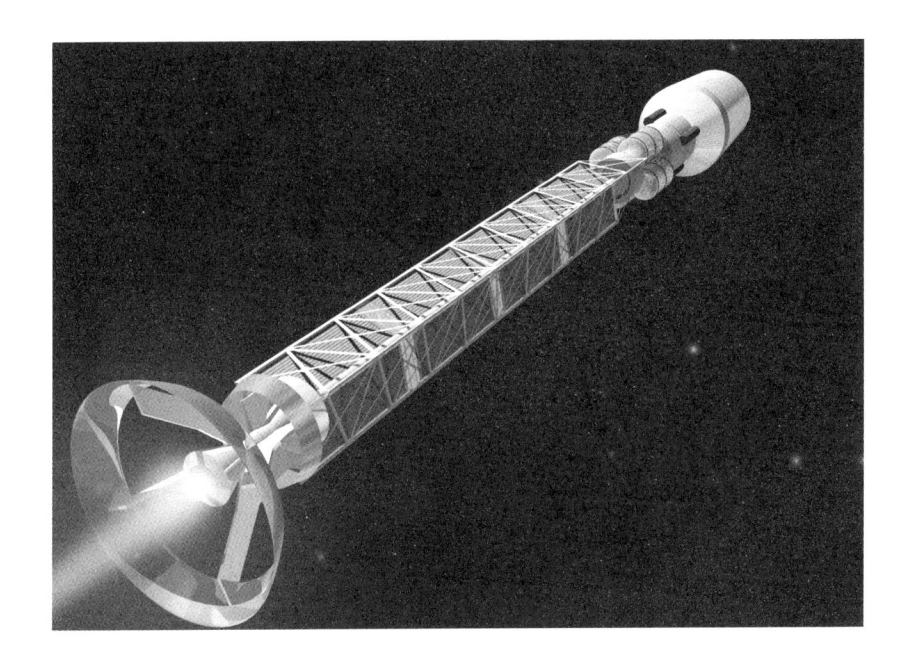

There in no time

Nothing can travel at light speed – except light, of course. So what would a journey across space be like for a photon, a single light quantum? Distances would shrink to zero and the photon's clock would stop ticking altogether. For the photon, there is no distance and no time; a journey from one side of the Universe to the other is accomplished in no time at all because for the photon the whole

Universe is contracted into zero length. The photon effectively is emitted and absorbed instantaneously. From the perspective of the photon it is as if the photon never existed, because what can exist for zero time? What all this actually means is something way beyond human comprehension, but as Einstein himself said:

'The most beautiful experience we can have is the mysterious. It is the fundamental emotion that stands at the cradle of true art and true science. Whoever does not know it and can no longer wonder, no longer marvel, is as good as dead, and his eyes are dimmed.'

Chapter 10

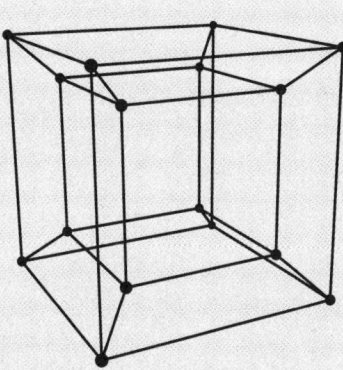

What is spacetime?

Einstein showed that even though space and time could be changed, the new concept of spacetime was absolute.

'Henceforth, space by itself, and time by itself, are doomed to fade away into mere shadows, and only a kind of union of the two will preserve an independent reality.' So wrote the German mathematician Hermann Minkowski (1864–1909), a year after Einstein published his special relativity theory.

Einstein argued that absolute time and absolute space could be consigned to the dustbin

'The ordinary adult never gives a thought to spacetime problems... I, on the contrary, developed so slowly that I did not begin to wonder about space and time until I was an adult. I then delved more deeply into the problem than any other adult or child would have done.'

Einstein wrote this to Nobel laureate James Franck. Einstein believed that it was usually children, not adults, who pondered spacetime problems.

and replaced by absolute spacetime. The mathematical reality of relativity shows that space and time are inextricably linked and both, as we have seen, are altered when we approach near-light speeds. Only by considering space and time together can we give an accurate description of what is observed at light speed.

The block Universe

To help visualize a path through spacetime, physicists employ a concept called the block Universe. Picture the Universe as a huge, rectangular box. Now try to imagine it as a four-dimensional box by adding time as the fourth dimension. A bit tricky isn't it? It's easier to simplify the picture by flattening space into two dimensions and swapping the third spatial dimension, going from left to right, for time. Taking a slice through the box gives us a snapshot of the block Universe at a moment in time. Any event at any place in our flat Universe can be mapped on the box, its coordinates showing us where it happened and when it happened. In fact, the spacetime block Universe maps all events, past, present and future.

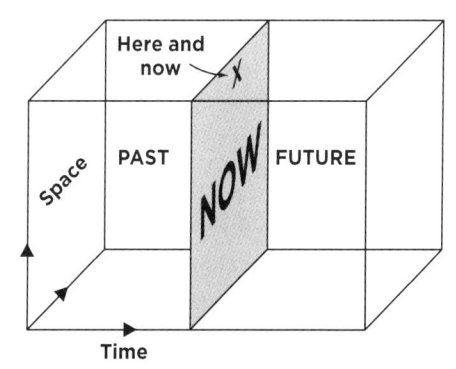

What is spacetime? | **111**

How does time flow through the block Universe from the present to the past? One view is that 'now' is the slice taken at this very moment. Time 'flows' in a series of infinitesimal jumps from slice to slice, each jump so small that we could never detect it. In another view, all of the slices exist simultaneously with all past and future events mapped along the timeline, although we are prevented from seeing this due to our inability to step outside the four dimensions of spacetime.

Whether time flickers into the future, like the pages of a flipbook, or whether the future already and unalterably exists, Einstein's relativity ties us into a picture of the universe in which space and time are inextricably linked. As we have discovered, because of the effects of time dilation and length contraction, spacetime divides into its space part and time part differently for observers in reference frames moving relative to each other. There is no way, for example, to claim unambiguously that an event lasted ten seconds without giving some indication of the reference frame in which the measurement was made. Observers in relative motion will not, in other words, be able to agree on which page of the flipbook an event took place.

Minkowski spacetime diagrams

In 1907, Hermann Minkowski developed another way of visualizing how objects moved through space and time. These representations in spacetime are called Minkowski spacetime diagrams and they give us a graphical way of visualizing some of the odd effects of relativity.

In a Minkowski spacetime diagram, a coordinate system is used, with time shown vertically on the y axis, and either one or two of the space dimensions represented along the x and z axes like a perspective drawing. If you want to think of a Minkowski diagram in a similar way to the block Universe then in this case the time slices are stacked vertically, with the past at the bottom of the pile. Each of these slices is called a spacelike hypersurface. In reality, these spacetime snapshots are three-dimensional, not flat surfaces, but as we saw with the block Universe, visualizing four-dimensional space is hard work!

In a Minkowski diagram, an object is not represented as a single point but as a line containing all the spacetime points at which it exists. This is the object's worldline. If it is in uniform motion, the object's worldline will be straight, but any force acting on it will cause the worldline to curve. If one object's worldline meets that of another object then the two objects collide at that point. The units along the time axis are usually given as seconds × the speed of light, so that the worldlines of light rays make a 45-degree angle with each axis.

The fact that nothing can travel faster than light places a restriction on how events can influence each other in spacetime. The path of all possible light-speed worldlines coming from an event spread out from it in a growing circle, like ripples spreading across a pond from where a fish has leapt up. Imagine all these split-second circles stacking one on top of the other up the timeline. As the circles, each one bigger than the one before, stack along the timeline they form an inverted cone shape, with its point at the origin of the event. This is called a light cone. The future light cone of the event maps out all the possible future events in spacetime that can be affected by the event. Because nothing can travel faster than light, anything outside the light cone cannot possibly be influenced by, or have any knowledge of, the event.

> **THE LAZY DOG**
>
> Albert Einstein was a student of Minkowski's at the polytechnic in Zurich. He didn't always impress Minkowski with his attitude to work. In conversation with Max Born about the theory of relativity, Minkowski remarked: 'It came as a tremendous surprise, for in his student days Einstein had been a lazy dog ... He never bothered about mathematics at all.'

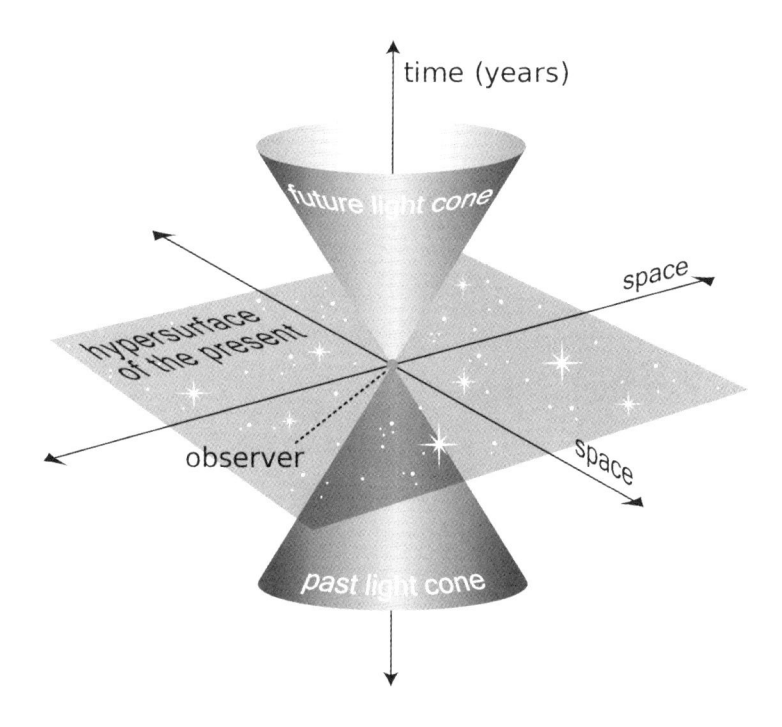

As well as the future light cone, we can also have an exactly symmetrical past light cone expanding out from the event into the past. The past and future light cones divide spacetime into three regions. The absolute future of the event is the region inside the future light cone. It contains everything that can possibly happen as a result of the event. The absolute past of the event is everything inside the past light cone. It contains everything that could possibly have caused or effected the event. Anything outside the past light cone can have had no effect on – or to have caused – the event in question. Everything that lies outside the past and future light cones of an event is said to be 'elsewhere'. Anything in the 'elsewhere' can have no knowledge of the event and can have no influence on it or be affected by it. Light cones are helpful because different observers will agree on the light cone of any event.

The British physicist Stephen Hawking gives the example of events that would take place on Earth if the Sun were suddenly to go out. Because of the time it takes the light from the Sun to reach Earth, we would know nothing about the extinguishing event until the Earth entered the Sun's future light cone eight minutes later. Until that time we would be quite unaffected by the fact that the Sun had gone out.

It isn't actually necessary to have light in order to have a light cone – not all events shed light after all. The light cone is simply a map of the geography of spacetime that shows the limits of the possible interactions that can be had with the event. Every event in spacetime has its own light cone so that spacetime is filled with an endless number of infinitely overlapping cones.

Spacetime and simultaneity

The spacetime diagrams can be used to explain some of the puzzling effects of special relativity, such as time dilation and length contraction. Any event that occurs simultaneously with an event on an observer's worldline will lie on a hypersurface that is perpendicular to that worldline. In other words, all points on the hypersurface lie at the same point in time, though they may be widely separated in space.

Now imagine a second observer, moving relative to the first. The second observer's worldline follows a path that is at an angle to the first's and so their hypersurface slice will also be tilted with respect to the first. This means that observer two cannot agree with observer one's notions of which events are simultaneous with each other.

It's worth reflecting on how important this is. Before Einstein shook things up in 1905, it was widely accepted that everyone experienced time in the same way, that we were all, as it were, on the same page in spacetime. In fact, we all have a slightly different version of the spacetime flipbook.

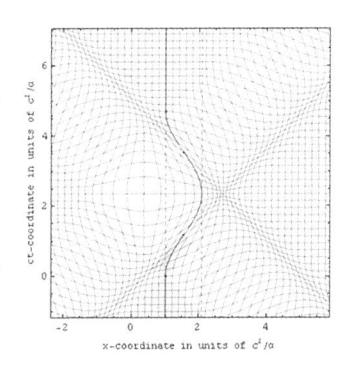

What is spacetime? | 117

Travelling through spacetime

Imagine you're in an aircraft flying due south. The pilot makes a course correction, with the result that the aircraft is now flying south-west. Now, the aircraft is still going in a southward direction but not as quickly as before because part of its velocity is now taking it west as well. What does this have to do with spacetime?

In the old Newtonian physics, travel through time and travel through space were held to be two quite separate things. But, according to Einstein, that's not the case. The two, as already stated, are inextricably linked. If you are stationary – that is, not moving through space – then all your spacetime movement is through time. When you start to move, some of your movement through time is diverted into movement through space. Like the plane changing course so that part of its velocity is taking it south, and part taking it west, the speed of your journey through time slows when some of your timelike motion is used for your journey through space.

Because the speed of light is constant, one observer's measurements of space and time differ from those of another observer in relative motion so that each measures the same value for the speed of light. According to special relativity, the combined speed of an object's motion through time and through space is precisely equal to the speed of light. This is an upper speed limit that can't be broken. For an object in motion, time must slow down otherwise the total combined speed through spacetime would exceed the speed of light. At the speed of light all the spacetime movement has become movement through space with nothing left over for movement through time. Which is why a photon of light, as we saw earlier, travels instantaneously across the Universe from its own perspective.

These relativistic effects on time are greater the closer we are to the speed of light but they apply for any movement through space, even at slow speeds. Experiments carried out using atomic clocks have demonstrated that clocks flown on a plane ran a few hundred billionths of a second slower than similar clocks that remained on the ground. It was a small difference to be sure, but it matched exactly with the predictions of special relativity.

Absolute spacetime

Einstein was never entirely happy with his theory being called 'relativity' theory. Indeed some things were, as he had demonstrated, relative, such as motion, distances and the duration of time, but they all took place against the unchanging backdrop of spacetime, the geometry of which was rigidly dictated by the speed of light. Absolute spacetime is as crucial to the understanding of special relativity as the absolute time and absolute space it replaced were to Newtonian physics. Einstein, in fact, preferred to think of his work as invariance theory, founded as it was on the unvarying nature of spacetime and the unchanging speed of light. Physicist Abraham Pais, who wrote probably the best scientific biography of Einstein, said there were two things that Einstein was particularly good at: 'He knew how to invent invariance principles and how to make use of statistical fluctuations.'

An invariant is something that stays constant under various transformations. A sphere is invariant, because it looks the same no matter how you turn it around. A cube, however, is invariant only under 90° rotations – if you turn it from face on to edge on it will look different. Einstein's insight in the special theory was that the speed of light is such an invariant. It is constant, no matter who measures it or how fast they are travelling at the time.

Chapter 11

Why does E = mc²?

It's one of the most familiar formulae in all of science, but what does it actually mean?

So far we've been looking at what happens to objects in uniform motion. As we have seen, as an object approaches light speed time for it slows down and its length contracts along the direction of travel. But what causes an object to be in motion? As Newton stated, an object will remain at rest or in uniform motion unless acted upon by a force. The changes Einstein's theory made to dynamics – the study of forces and motion – led to the famous equation $E = mc^2$.

Einstein published his idea in a short paper, just three pages long, that was a sort of coda to the special theory of relativity. It was entitled 'Does the Inertia of a Body Depend Upon its Energy-Content?' and was submitted to the *Annals of Physics* in September 1905.

Force and momentum

An object in motion is said to have momentum, which is a measure of the quantity of motion. It is defined by the equation: momentum = mass × velocity. Increasing either the mass or the velocity of an object increases its momentum. When two moving objects collide, energy and momentum are transferred between them. This exchange of energy and momentum results in a force acting on the two objects. A force is a measure of the rate of transfer of energy and momentum. Force, energy and momentum are related to each other by these formulae:

WHAT IS ENERGY?

The E in the equation stands for energy, but what is energy? The modern use of the term dates from around the 1840s when it was first used by physicist William Thomson (who later became Lord Kelvin). He realized that the power that drove many different processes could be explained in terms of the transfer of energy from one system and form to another. Energy comes in different forms. For example, there's the chemical energy stored in your muscles that allows you to move, there's kinetic energy, the energy of movement, there's potential energy, such as the energy stored in a tightly drawn bow string, there's electromagnetic energy, heat energy, and atomic energy. Energy makes things happen; without energy nothing would happen. The more energy that there is available the more that can be achieved. If an object is described as energetic it means it can do things. Scientists believe that the amount of energy in the Universe is limited – energy can change from one form to another but it cannot be created or destroyed.

Momentum gained = force × time during which the force acts.
Energy gained = force × distance through which the force acts.

These formulae are true in both classical Newtonian physics and in Einstein's relativistic physics.

Just as energy is conserved in any interaction, the total amount of energy present being the same at the end as it was in the beginning, so too is momentum conserved. Newton's third law of motion, 'for every action there is an equal and opposite reaction', arises from this. When a rocket is launched, for example, the upward momentum of the rocket is balanced by the downward momentum of the hot gases expelled from its engines.

In classical physics, it was theoretically possible to impart as much momentum to an object as you wanted and accelerate to any speed you desired; all you had to do was apply a large enough force for a long enough time and it should even be possible to exceed the speed of light. This, of course, is something that relativity theory simply does not allow.

In relativistic physics, it is also possible to impart unlimited momentum to an object by applying a force to it. But, no matter how much or for how long a force is applied, the object will never accelerate beyond the speed of light. In the classical framework it was naturally assumed that the mass of the object stayed the same

WHAT IS MASS?

Perhaps the simplest way to put it is to say that mass is the amount of 'stuff' something contains. Mass isn't the same as weight. You can measure mass by measuring weight – the gravitational mass. A 10 kg bag of potatoes is definitely more massive than a 5 kg bag, but the answer you get will depend on how strong the force of gravity is where you do the weighing. Your 10-kg sack of potatoes will only weigh about 1.6 kg on the Moon for example, but you will still have the same amount of potatoes: that is, their mass will be the same. Mass is also a measure of the amount of inertia, or resistance to movement, an object has – its inertial mass. From Newton's force = mass × acceleration equation (F = ma) you can also determine how massive an object is from the amount of force you have to apply to get it moving.

and an increase in momentum meant an increase in velocity. Einstein, however, decreed that this wasn't the case at all – as the velocity of an object increases, so too does its mass. As the object approaches light speed less and less of the increase in momentum is taken up by an increase in velocity and more and more by an increase in mass.

Just like the effects of time dilation and length contraction earlier, the mass increase isn't felt by the object itself. The crew of a spacecraft approaching light speed wouldn't find themselves getting heavier and heavier. The increase in mass would only be apparent to an external observer who was relatively stationary compared to the ship and sees that it is resisting being accelerated.

According to relativity theory, an object cannot be accelerated beyond the speed of light, because the closer it gets to the speed of light, its mass increases exponentially towards infinity so it becomes harder and harder to accelerate it as the amount of energy required to do so also increases. When Einstein published his theory, it was known that it grew increasingly more difficult to accelerate electrons in a cathode ray tube as they approached light speed. At the time, it was thought that this was caused by some relationship between the electron and the electromagnetic field, but Einstein showed it was the result of an increase in the electron's mass.

Kinetic energy

The energy of movement, kinetic energy, is given by the equation:

$E = 1/2mv^2$

This means that the kinetic energy of an object is equal to half its mass multiplied by the velocity squared. This is fine for everyday 'low' speeds but it is going to become increasingly inaccurate approaching light speed as m, the mass, begins to increase.

A moving object increases in mass and has kinetic energy by virtue of its motion. As a moving object slows down it loses kinetic energy. An object at rest has zero kinetic energy. The mass of an object cannot be zero of course. The lowest mass an object can have is called its rest mass and its mass while in motion is its relativistic mass.

Finally, E = mc²

We're just about ready to put it all together into the famous equation. If an object is moving at very close to the speed of light then, as we have seen, any force acting on it, and imparting energy and momentum, will cause it to increase in mass because it can't go any faster. Going back to the relationships between force, energy and momentum, we saw that energy gained equals the force multiplied by the distance through which the force acts. Since the object is travelling at a speed very close to that of light, the distance it moves is as near as makes no difference the same as that travelled by light in the same time. So we can write this as an equation:

$E = \text{force} \times c$

(where E = energy and c = the speed of light)

From the second relationship, momentum gained equals the force multiplied by the time during which the force acts, we can also write a second equation. As momentum equals mass × velocity and the velocity doesn't change during the time in which the force acts, the mass increases by an amount and the velocity remains at close to the speed of light. We can represent this as:

Force = m × c

(where m = mass)

These two equations can be combined into:

E = force × c = (m × c) × c

And this can be simplified into:

$E = mc^2$

Two sides of the same coin

Einstein's equation effectively says that energy and mass are the same thing. If an object gains or loses mass or energy it gains or loses an equivalent amount of energy or mass in accordance with $E = mc^2$. So far we've only explored this in terms of the energy of motion, kinetic energy, but does it hold true for other forms of energy? For example, does an object lose mass as it cools down? Actually, yes it does. After all, temperature is a measurement of how rapidly the atoms and molecules that make up a substance are moving around, so in accordance with $E = mc^2$, the faster they move they more massive they'll be.

Einstein believed that his formula would explain a curious discovery that had been made by the Polish physicist Marie Curie. She had observed that an ounce of radioactive radium would produce 4,000 calories of heat per hour, seemingly indefinitely. Where, she wondered, was this energy coming from? According to Einstein, as the radium radiated heat it should also be losing mass. Unfortunately, the equipment available at the time wasn't accurate enough to measure the tiny amount of mass that was being converted into energy and there was no way to experimentally verify Einstein's explanation. Einstein wrote: 'The idea is amusing and enticing, but whether the Almighty is laughing at it and leading me up the garden path – that I cannot know.'

Years later, in 1948, Einstein explained the equivalence of mass and energy like this:

'It followed from the special theory of relativity that mass and energy are both but different manifestations of the same... Furthermore, the equation $E = mc^2$, in which energy is put equal to mass, multiplied by the square of the velocity of light, showed that very small amounts of mass may be converted into a very large amount of energy and vice versa. The mass and energy were in fact equivalent, according to the formula mentioned before.'

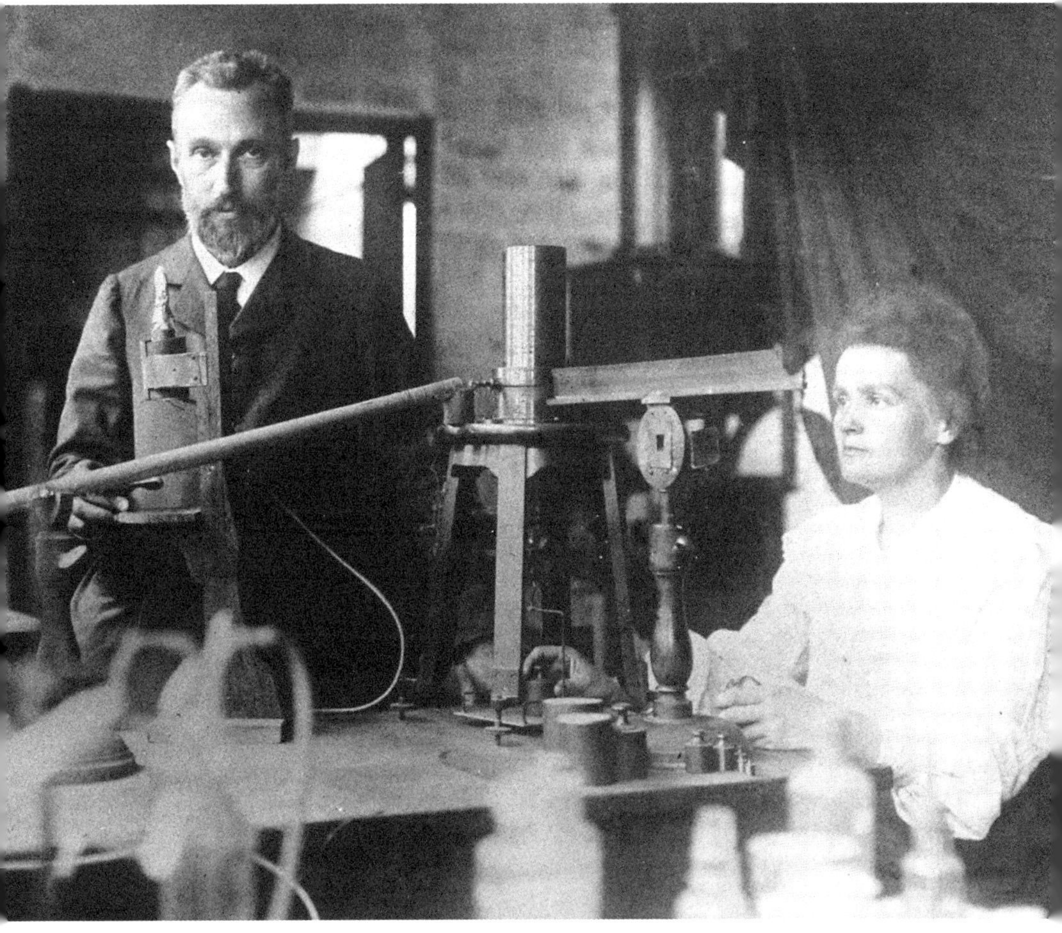

The speed of light is a very big number. Squared it is a very big number indeed. This means that if it were possible to convert even a tiny amount of matter into energy the output would be enormous. Physicist Richard Wolfson has calculated that there is almost enough energy stored in a raisin to provide all of New York City's needs for a day.

Chapter 12

How did Einstein fit gravity into relativity?

Special relativity was only the beginning, now Einstein had to find a way of bringing gravity into his calculations.

This was the question Einstein set out to answer in the general theory of relativity, published in 1915. In setting out the special theory, Einstein had concentrated solely on objects moving with uniform motion. He had chosen to ignore objects that were accelerating and objects affected by gravity. He did this for the very good reason that it made the calculations a lot easier. Writing in *The Times* of London newspaper on 28 November 1919, Einstein said:

'…the theory of relativity resembles a building consisting of two separate storeys, the special theory and the general theory. The special theory, on which the general theory rests, applies to all physical phenomena with the exception of gravitation; the general theory provides the law of gravitation and its relations to the other forces of nature.'

Formulating the general theory would take Einstein seven years of sometimes very intense work. Physicist Dennis Overbye described Einstein's achievement as 'arguably the most prodigious effort of sustained brilliance on the part of one man in the history of physics'. What emerged would be a view of the Universe that was totally unlike any that had gone before. Once again, Einstein would change our whole conception of the way the Universe worked.

The gravity conundrum

Einstein faced a conundrum. Special relativity was built around the fact that light was the fastest thing in the Universe. This directly contradicted Newton's ideas about how gravity worked. According to Newton, gravity made its effects felt instantaneously, from the Sun holding the Earth in orbit, to the path of a space probe to Pluto or a skydiver falling back to Earth, gravity acted without delay, apparently

How did Einstein fit gravity into relativity? | 137

propagating through space faster than light. If the Sun were to disappear, the Earth would slingshot out of its orbit in the same second. No waiting eight minutes for it to enter the Sun's light cone. If Einstein was right, then that couldn't be possible.

Possible or not, Newton's universal law of gravity had been backed up by observation and experiment time and time again. So how did gravity make its influence felt? Obviously, it was a force that acted at a distance and didn't require any physical contact to work. Also, unlike any other force, it was impossible to shield yourself from its effects. Newton's law explained how to calculate the effects of gravity but he didn't even try to explain what caused it. Passing the buck somewhat he wrote in the *Principia*: 'I leave this problem to the consideration of the reader.'

Acceleration

As Einstein had shown, if you are in uniform motion then it's impossible to demonstrate that you actually are moving. All observers moving uniformly relative to each other are entitled to say that they are stationary and that it's everyone else who is moving.

Accelerated movement is quite different. If we change speed or direction, we feel it. Without looking out of the window, you know when the train is going round a bend because you feel yourself leaning sideways, when an aircraft begins to accelerate down the runway for take-off you feel yourself pressed into your seat, with no visible clues at all you can tell when a lift has started to move up or down. When we accelerate, we feel inertial forces – the forces that resist a change in speed or direction. These are the forces that throw us against the side of the train when it takes the bend; and they are the forces that cause our coffee to part company with its cup when the bus we are travelling in hits a pothole.

One, two, free fall...

In his legendary experiment at the tower of Pisa, Galileo demonstrated that a small stone will fall to the ground in the same time it takes

How did Einstein fit gravity into relativity?

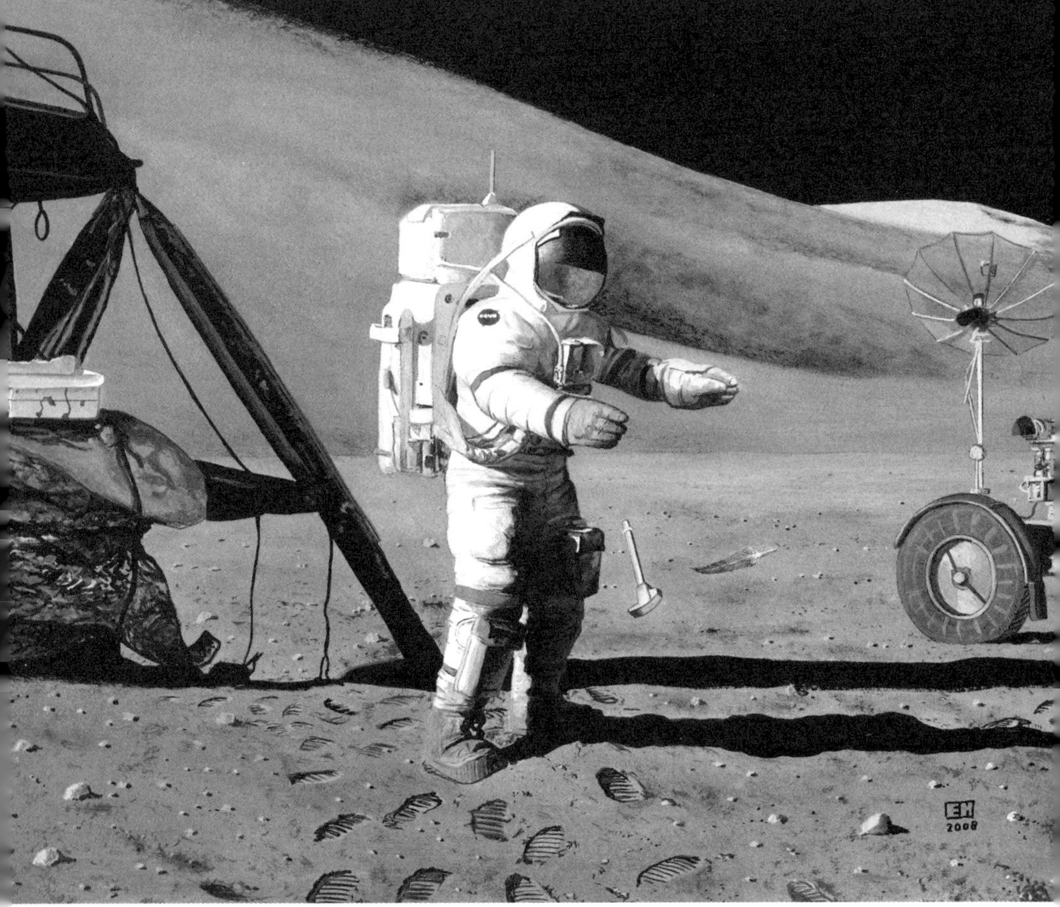
EH
2008

a large stone to do so. This is because the two stones accelerate towards the ground at the same rate and this holds true no matter what the difference in their mass. Galileo couldn't explain why this was so but Newton could with his second law of motion – force equals mass times acceleration. Plug the appropriate numbers in and the answer for the acceleration of a falling object comes out the same – 9.8 mps on Earth (other planets differ). This idea, that gravity accelerates all objects at the same rate regardless of what they might be made of, is called the 'Universality of Free Fall' or the 'Equivalence Principle'. Einstein would build his theory of gravity by assuming the Equivalence Principle to be true.

This effect came about because two quantities in Newtonian theory, the inertial mass of a body and its gravitational mass, matched exactly. Einstein believed that this couldn't be coincidence. If he was going to find a workable theory of gravity it would have to explain this phenomenon.

Einstein's happiest thought

In what he called his 'happiest thought', which he most likely had some time in November 1907, Einstein realized that gravity and acceleration were equivalent – without a frame of reference you can't tell one from the other. Lecturing in Kyoto, Japan, in 1922 he said: 'I was sitting in a chair in the patent office at Bern when all of a sudden a thought occurred to me: If a person falls freely he will not feel his own weight. I was startled. This simple thought made a deep impression on me. It impelled me towards a theory of gravitation.'

Should you have the extreme misfortune to be in a lift on the top floor of a skyscraper when the cable snaps you will immediately begin plunging to your doom. However, you should have time to observe some intriguing phenomena. Your feet are no longer pushing against the floor of the lift. Should you choose to jump off the floor you will not fall back. It is as if gravity has vanished. There are no experiments that you can perform that will conclusively show whether you are heading towards the ground or floating freely in deep space far from any gravitational force. Close your eyes (you probably will) and you can imagine yourself to be a weightless astronaut (which might be comforting under the circumstances). In both cases the laws of physics are the same.

Einstein in a box

In another of his famous thought experiments, Einstein developed the idea something like this. Imagine a physicist who awakens in a box. Unknown to the physicist, the box is no longer on Earth but in deep space and is under uniform acceleration. If the physicist were to release objects in the box, their inertia would cause them to fall to the 'bottom' of the box, that is, in the opposite direction to that in which the box is moving. All objects dropped by the physicist would fall exactly alike, no matter what their mass or composition, in accordance with Galileo and Newton. The physicist would conclude from this observation that there was a gravitational field at work inside the box.

Einstein's assertion was that this wasn't just an effect that was similar to a gravitational field, it actually was a gravitational field. He formulated a principle of equivalence that stated that the effects of uniform acceleration were indistinguishable from the effects of gravity. Acceleration creates a gravitational field. According to Einstein's equivalence principle, whether the physicist in the box was accelerating or not depended on your point of view. The physicist in the box would consider himself to be in a gravitational field and not accelerating, but an observer watching the box would see it accelerating uniformly through gravity-free space. Each relative point of view is equally valid. This was what made the inertial mass and the gravitational mass the same.

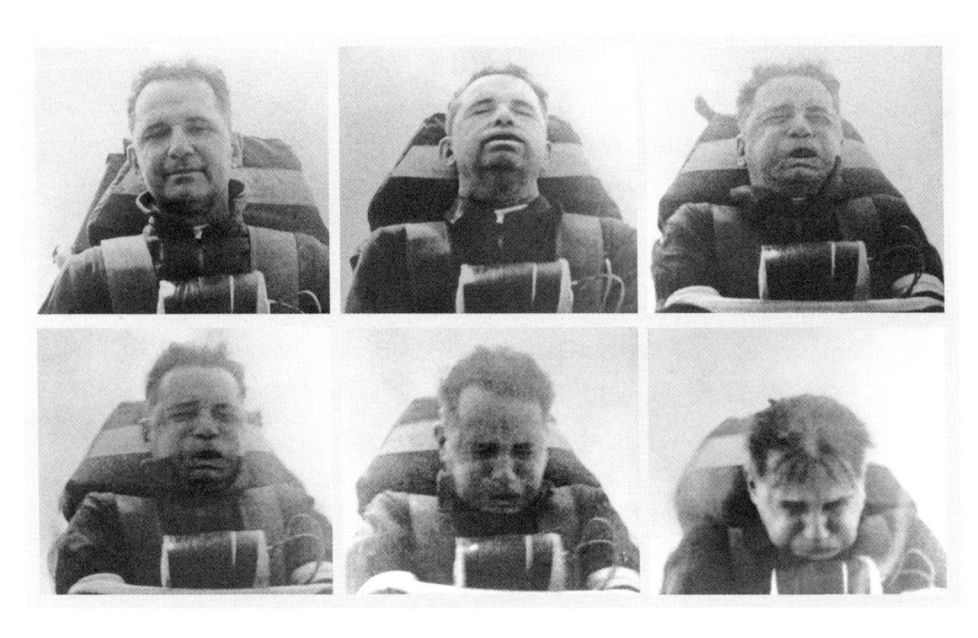

THE VOMIT COMET

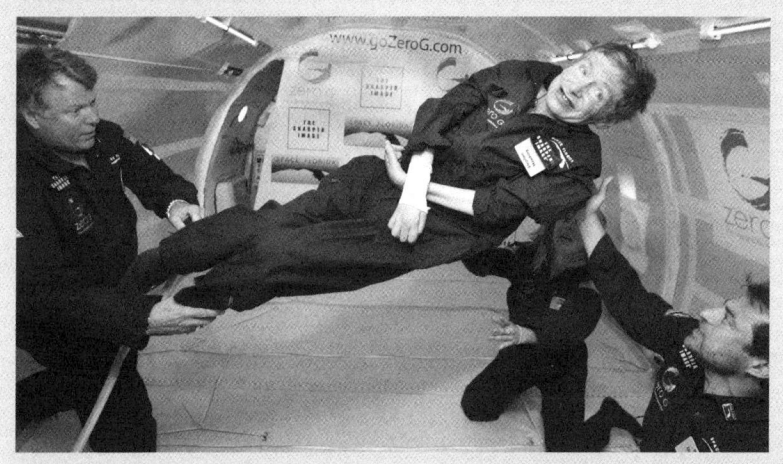

NASA trains its astronauts aboard a modified aircraft that flies on a parabola-shaped free-fall flight path that allows the passengers on board to experience near weightlessness for around 20 seconds. The plane is known as the 'Weightless Wonder' or the 'Vomit Comet' as the effects of weightlessness can induce nausea in some people. The aircraft can fly 40 to 60 parabolas in the course of a research flight. First, the pilot takes the aircraft up at a 45-degree angle before throttling back the engines to slow the aircraft and pitch the nose down to complete the parabola. As the aircraft pulls out of its dive, the passengers and crew actually feel a force equivalent to twice that of normal gravity.

Redshift

Einstein's equivalence principle predicts that the wavelength of electromagnetic radiation will lengthen as it climbs out of a gravity well, a phenomenon called gravitational redshift. Thanks to Einstein's $E = mc^2$, and Planck's $E = \hbar f$ law relating the energy of light to its frequency, it becomes apparent that as a photon moves out of a gravitational field, it must lose energy. Since photons always travel at the speed of light, this energy loss is seen as a lowering in frequency, rather than as a reduction in velocity. This lowering of the frequency of the photon corresponds to a 'redshift' to the lower-frequency, longer-wavelength end of the spectrum.

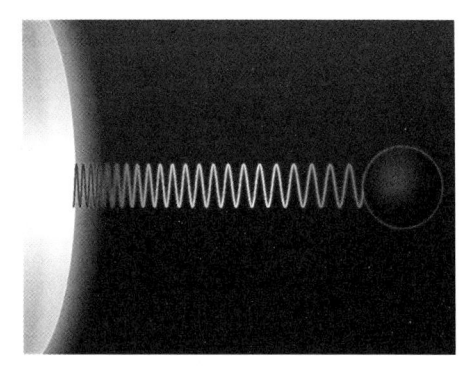

Another consequence of this, which may not be obvious, is the slowing of time. If we send a beam of light from the surface of the Earth to an observer high above it he or she will, as said, see its frequency go down, which means that the length of time between one wave crest and the next increases. It would appear to the high-flying observer that things down below were taking a little longer to happen. This prediction of general relativity was tested in 1962 when

LIGHT BENDS, TIME SLOWS

Einstein realized that one consequence of the equivalence principle is that the path of a light beam will be bent by gravity. Imagine a photon crossing the physicist's box as it accelerates through space. As the photon is crossing the box, the floor is accelerating upwards, which means that the photon appears to fall downwards. Because a gravitational field is equivalent to acceleration, the same must also hold true there.

A second consequence is that time slows in a gravitational field. This effect, called gravitational time dilation, means that observers at different distances from a large object (which produces a gravitational field) will obtain different measurements for the time elapsed between two events. This is a direct consequence of the fact that an observer outside the box, that is, one outside the gravitational field, sees the photon follow a straight path, but the physicist in the box sees it follow a longer, curved path. Because the speed of light cannot change, the physicist's clock has to run slower to allow both journeys to be made in the same time.

How did Einstein fit gravity into relativity?

two extremely accurate atomic clocks were placed on a tower, one at the top and one at the bottom. The clock at the bottom, the one deepest in the Earth's gravity well, ran slower than the one at the top. The discrepancy was exactly in line with prediction.

A second example of redshift, called the cosmological redshift, appears to be evidence that the Universe is expanding. We will return to that later.

Solving the twins paradox

Earlier, we looked at the paradox of the twin in the spaceship travelling at close to light speed who returns home to discover that she was now younger than her sibling who stayed behind. The truth is that although, relatively speaking, it could be argued that it's just as valid to say the space station retreated from the spacecraft as vice versa, in fact the situation wasn't the same for both twins. The twin in the spaceship had to accelerate to get close to light speed (and decelerate to slow down, turn around and come back). Acceleration is equivalent to gravity and gravity, as we have seen, slows time. It follows, therefore, that acceleration slows time too. The reason the spacefaring twin ages less than her sibling is because she was accelerating while her sibling was not. Their situations were not symmetrical after all. Once again, this demonstrates that there is no absolute time in relativity; time is something personal to everyone, measured according to where we are and how we are moving.

To the equivalence principle, Einstein added the relativity principle, which stated that the laws of physics in any frame of reference are governed by the theory of special relativity. These were the foundations upon which he built his theory of general relativity. This would extend the concept of spacetime forged in the special theory of relativity to all of physics and, particularly, to the theory of gravitation.

How did Einstein fit gravity into relativity?

Chapter 13

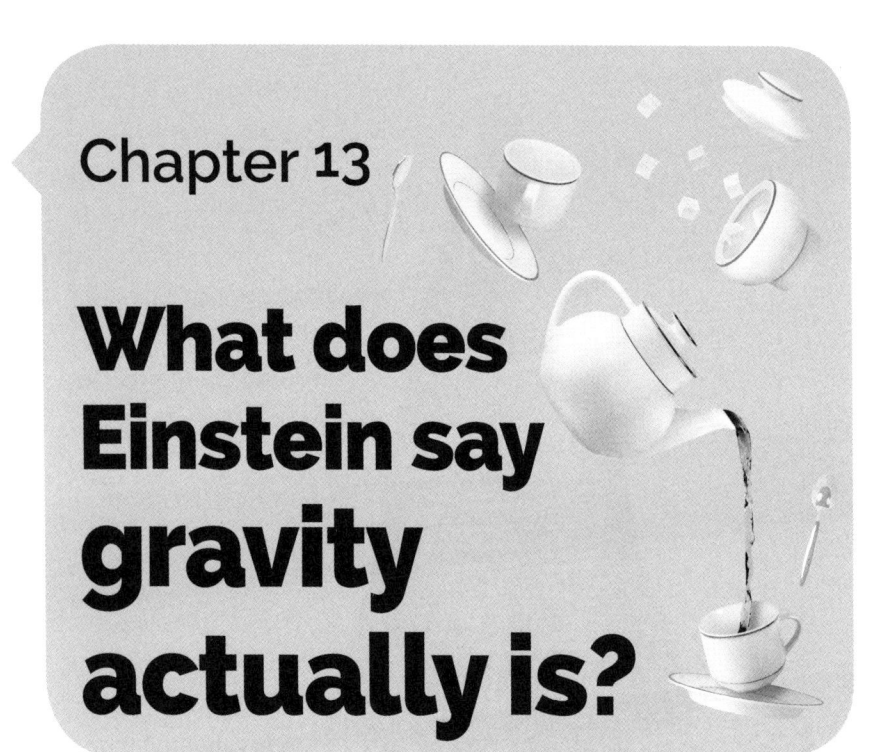

What does Einstein say gravity actually is?

For centuries gravity was believed to be a force of attraction between two objects, but to Einstein it was a distortion in the very fabric of spacetime.

What does gravity do to spacetime? The key idea in general relativity is that gravity is not a force acting between objects but rather the result of a distortion of spacetime caused by the objects in it. The bigger the object, the more spacetime curves around it.

Gravity and tidal forces

Imagine two spacecraft moving along parallel trajectories through empty space at the same speed relative to each other. So long as no force acts on them, their paths will follow a straight line. Now suppose there's a planet up ahead. According to Newton, its gravity exerts a force that will tug the spacecraft off course, causing their paths to converge. This happens because they are both being pulled towards the planet's centre of gravity, so they are both moving towards the same point in space. It is this difference in direction that's responsible for the distance decreasing between the two spacecraft. Physicists call this force difference a 'tidal force'. It is so-called because it is just such a difference between the Moon's gravitational attraction on the Earth and the Earth's oceans that causes the tides to rise and fall.

Tidal forces also show us that gravity is not entirely 'switched off', even in free fall. For a human-sized object free-falling in Earth's gravity there will be a tidal force acting because your feet, closest to the ground, are feeling a very slightly stronger gravitational pull than

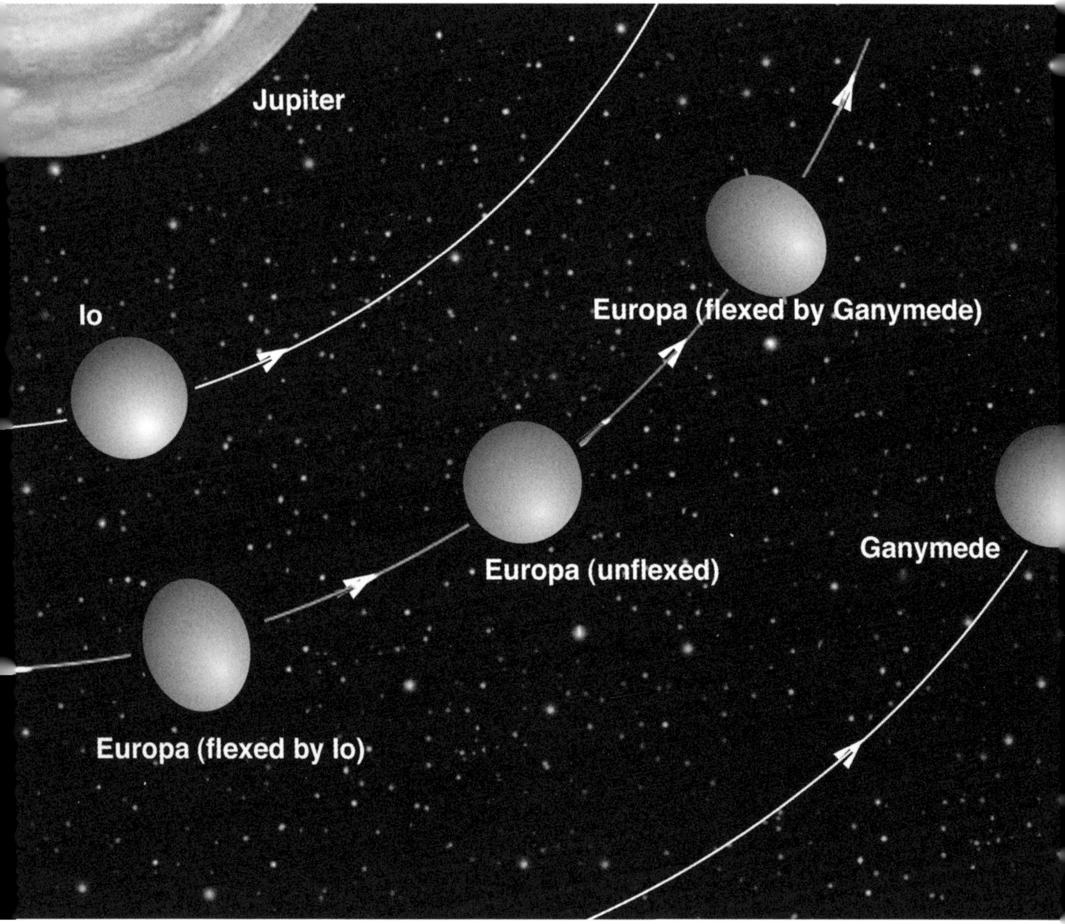
Jupiter
Io
Europa (flexed by Ganymede)
Europa (unflexed)
Ganymede
Europa (flexed by Io)

your head, which is further away from the centre of gravity. It's an insignificantly small difference but it is there.

Curved surfaces

We are used to dealing with the geometry of flat plane surfaces. It's what we learned in school. The sum of all the angles in a triangle is always 180 degrees; two parallel lines will never meet and so on. But there is a way two parallel paths can converge.

The surface of a sphere is also a flat surface, but this time it is curved and the laws of geometry become a little different. There are no straight lines on a sphere, or indeed any type of curved surface, but we can construct lines that are as straight as possible. Mathematicians call these straight-as-possible-lines 'geodesics'. In the case of the sphere, the shortest distance between two points will lie along the path of a great circle, the largest circles that can be drawn on that surface. Great circle routes are the ones taken by commercial airline pilots to ensure they follow the shortest flight path between two airports.

Now, instead of spacecraft, imagine two supersonic skimmers heading north in parallel from the equator to the North Pole. Without changing course, they will collide at the North Pole when their paths meet. No force has acted on them but because they were moving on the curved surface of a sphere their courses inevitably cross. To see why this happens we can construct a triangle on the surface of the sphere from three intersecting geodesics. The bottom of the triangle lies on the equator and represents the distance between the two skimmers. The two angles on the equator are both right angles, so the initial flight paths are parallel. But the paths converge on the North Pole tip of the triangle. You may also have noticed incidentally that the angles of the triangle are going to add up to more than 180 degrees.

It is only when you look at larger regions of the sphere that it becomes apparent that the surface is curved. You could walk all day and not get any impression at all that you were moving along the surface of a sphere, but rise a dozen kilometres or so into the air and you will just be able to discern the curvature of the Earth. The same is true for any curved surface: a small enough section of the surface is effectively indistinguishable from a flat plane.

Curved spacetime

Einstein took this property of curvature and drew a comparison with the way gravity works. For a very small region of spacetime, for instance a physicist floating freely in a box in space moving uniformly, there is no gravity. The inside of the box obeys the laws of the spacetime of special relativity. The spacetime of special relativity, where gravity is absent, is analogous to the flat surface and the laws governing motion are fairly straightforward. So long as no force acts on an object it will continue to move in a straight line at a constant velocity, following a straight path through spacetime.

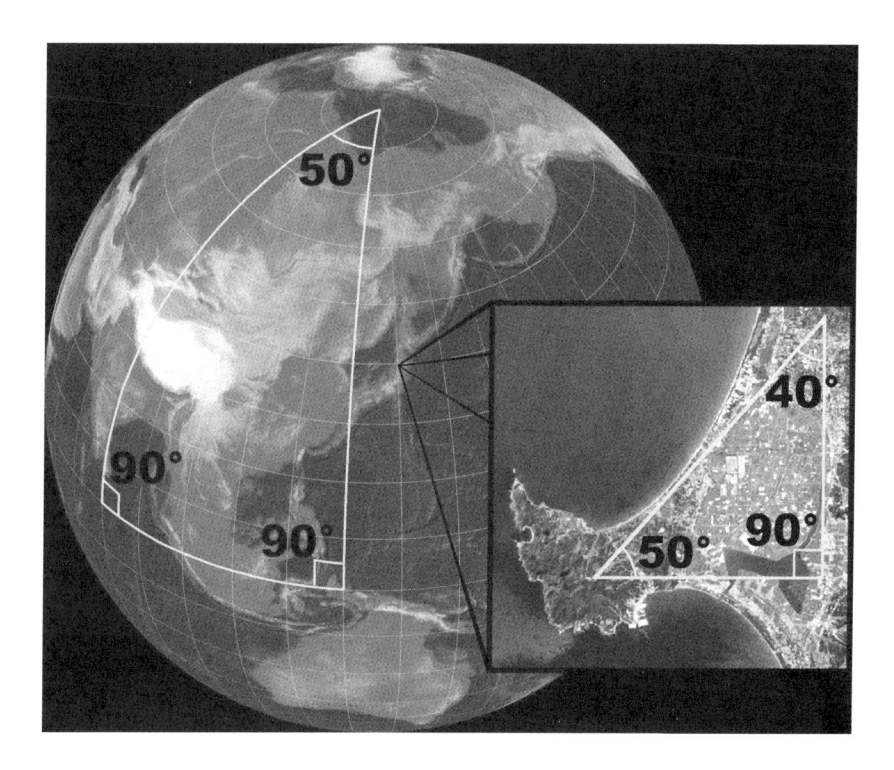

If we now add gravity to the situation, for instance by placing a large planet in the path of the physicist, Newton dictates that the

What does Einstein say gravity actually is? | **155**

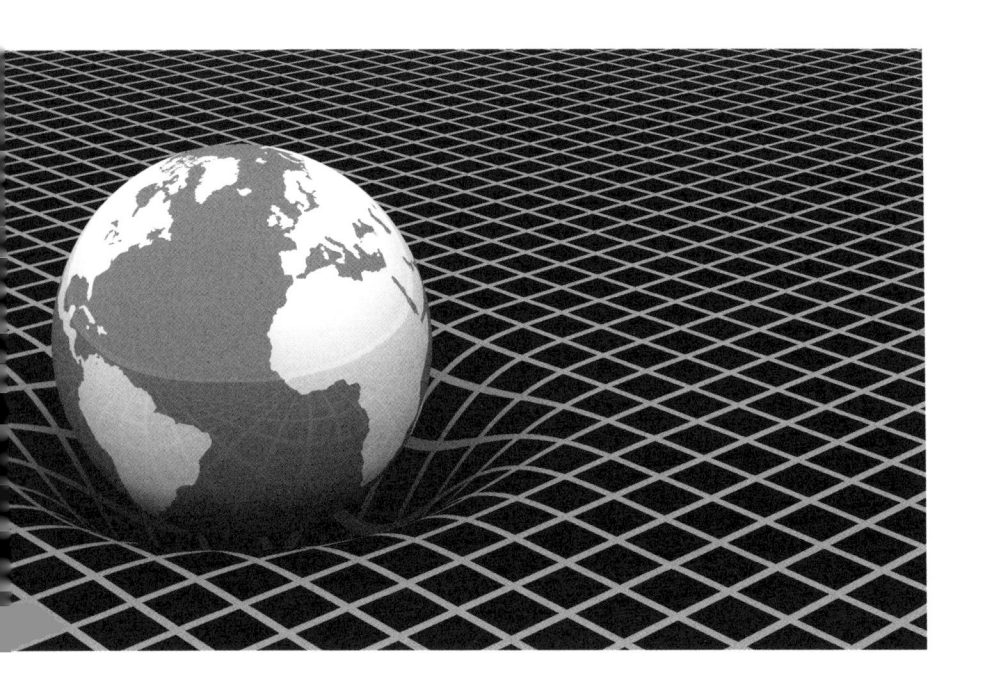

planet will exert a force on all objects around it. The hapless physicist will begin to feel the first effects of acceleration as his path begins to curve towards the planet.

In his theory of gravity, Einstein looks at things in a quite different way. Rather than exerting a force, a mass causes a distortion of spacetime. Empty spacetime, the spacetime of special relativity, is flat. But where there is matter present, spacetime is curved. In the same way that there are no straight lines on the surface of a sphere, there are no straight lines in curved spacetime. The closest we can get to the straight line in curved spacetime, just as on a sphere, is a geodesic, a curve that is as straight as possible. The physicist heading towards the planet has not been deflected from his straight-line course, rather the presence of the large planet distorting spacetime has changed the form a straight line can take. It has redefined the geometry of spacetime. According to general relativity, an object follows a straight line geodesic through spacetime, but from our three-dimensional perspective the path looks curved!

The Earth makes a dent in spacetime, curving it around itself. The Moon follows a straight path through the Earth-curved spacetime, which to us appears to take it on a circular orbit (elliptical really as the Moon curves spacetime too) around the Earth. General relativity predicts that light rays will be bent by gravitational fields because light also follows geodesics through spacetime. This bending of light by gravity was, as we shall see, one of the first confirmations that Einstein's theory was right.

A dance to the music of spacetime

This is the basis of Einstein's theory. Newton's gravity is a force that acts on objects and influences their movement, but gravity, in Einstein's universe, is the result of curved spacetime, a distortion

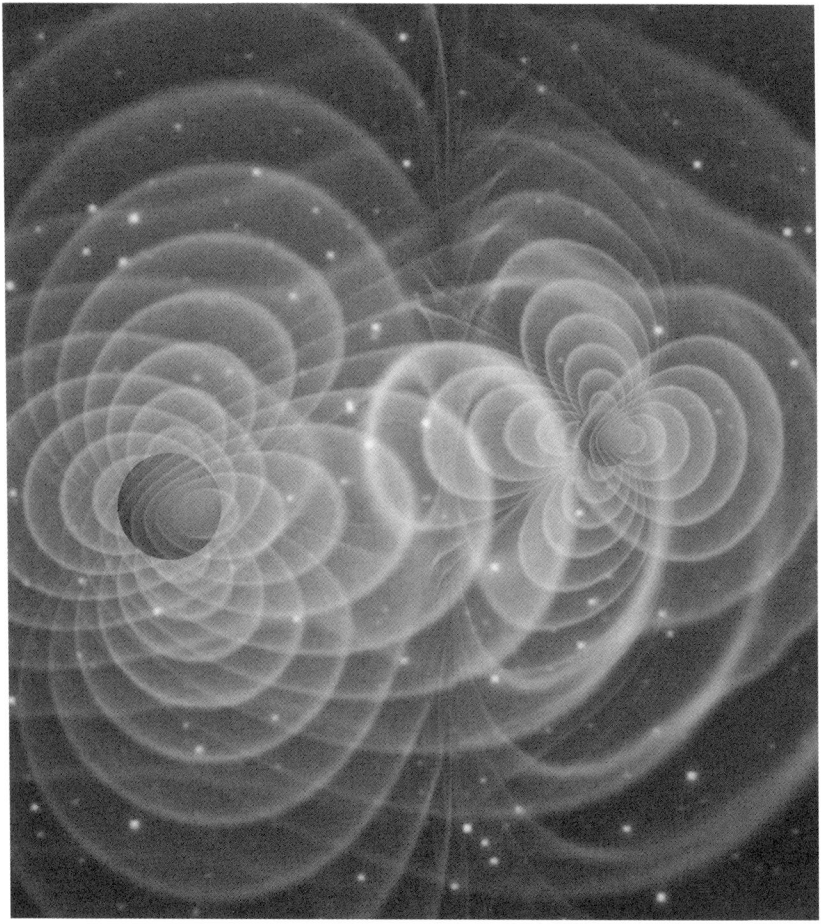

of spacetime geometry. Objects still follow the straightest possible paths through spacetime, but because spacetime is now curved, they accelerate as if they were under the influence of a gravitational force.

In Einstein's universe, matter and spacetime interact in a complex and ever-changing dance. Matter distorts the geometry of spacetime and this distorted geometry dictates how matter moves through it. As the matter moves and the sources of gravity change positions, so the swirling curves of spacetime ebb and flow also. As physicist John Archibald Wheeler succinctly summarized it: 'Spacetime tells matter how to move; matter tells spacetime how to curve.'

Gravitational waves

One of the predictions of general relativity was that there should be a phenomenon called 'gravitational waves'. Gravitational waves are like ripples in spacetime caused by particularly energetic disturbances. Einstein's equations showed that cataclysmic events, such as two black holes colliding or a massive supernova explosion, would be like large rocks being dropped into the pond of spacetime, sending out waves of distorted space across the Universe at the speed of light.

Although gravitational waves were predicted to exist in 1916, there was no actual proof of their existence until 20 years after Einstein's death. In 1974, astronomers at the Arecibo Radio Observatory in

Puerto Rico discovered a binary pulsar – two extremely dense and heavy stars in orbit around each other. Knowing that this system could be used to test Einstein's prediction, the astronomers began making careful observations of the system. Eight years of meticulous data-gathering revealed that the pulsars were getting closer to each other at exactly the rate general relativity predicted that they would. After over 40 years of close monitoring, the observed changes in the orbits of the pulsars is in such close agreement with general relativity that the researchers have no doubt that it is emitting gravitational waves.

Until September 2015, all the confirmations of the existence of gravitational waves had been indirect or determined mathematically, and not through actual physical proof. On 14 September, the Laser Interferometer Gravitational-Wave Observatory (LIGO) here on Earth detected gravitational waves for the first time. The waves it detected were generated by two colliding black holes nearly 1.3 billion light years away! LIGO is incredibly sensitive. They may be generated by extremely violent events, but by the time the waves reach the Earth they are many millions of times smaller. In fact, by the time the gravitational waves detected by LIGO reached Earth the degree of spacetime wobbling they generated was much smaller than the nucleus of an atom, which may explain why you didn't feel them passing through.

LIGO is an absolute triumph of engineering skill and ingenuity. It consists of two L-shaped detectors built 3,000 km apart and housed inside vacuum chambers 4 km long. Working in unison, they can measure a motion 10,000 times smaller than an atomic nucleus – no measurement of this accuracy had ever been attempted before. It is equivalent to measuring the distance to the nearest star to a precision smaller than the width of a human hair.

Chapter 14

How did an eclipse prove Einstein was right?

Arthur Eddington's astronomical observations during a total eclipse of the sun confirmed that Einstein's relativity equations were correct.

In the autumn of 1919, Pauline Einstein received a postcard from her son, Albert. It began, 'Dear Mother, joyous news today. H. A. Lorentz telegraphed that the English expeditions have actually demonstrated the deflection of light from the Sun.'

When Einstein first set out his equivalence principle in 1907 and deduced that it would result in the bending of light he thought that the effect would be far too small ever to be measured. Einstein's first predictions for the bending of light from a star by the Sun were in accord with what Newton himself would have predicted from his law of gravitation and his belief that light took the form of a stream of particles. Einstein's answer was incorrect.

At this time, Einstein had not yet arrived at the insight that spacetime was curved and that this would have an effect on the bending of the light beam. It wasn't until 1915 that he realized that, according to his general relativity theory, the light ray passing the Sun would be bent by twice the value of that initial 1907 calculation. It is perhaps fortunate that there had been no opportunity to test Einstein's

How did an eclipse prove Einstein was right? | **163**

idea before he had made this correction, though he had a couple of lucky escapes. An expedition to view an eclipse in Brazil in 1912 had included measuring the deflection of light in its list of experiments but was prevented from doing so by bad weather. In the summer of 1914, a second expedition set off for the Crimea to observe an eclipse but was forced to turn back when the First World War broke out.

Now, with a clear difference between Einstein's predictions from general relativity and those of Newtonian physics, it would be possible to find out who was right, but it would have to wait until the war had ended before further observations could be carried out. Some attempts were made to find evidence of the predicted deflections in photographs of earlier eclipses but without success. Einstein was anxious to have his theory proved right. In a book he wrote in 1916 to explain relativity to a wider audience, he stated:

'The examination of the correctness or otherwise of this deduction is a problem of the greatest importance, the early solution of which is to be expected of astronomers.'

He would have to wait until September 1919 when two British expeditions finally obtained the results he had been hoping for.

The eclipse expeditions of 1919

Astronomer Sir Arthur Eddington led an eclipse expedition to the island of Principe, off the coast of West Africa, during the total eclipse of 29 May 1919. A second expedition, to Sobral in Brazil, was led by Andrew Crommelin from the Greenwich Observatory.

Eddington had been fortunate enough to obtain a copy of Einstein's theory in 1916, even in the middle of the war. He became an enthusiastic champion of relativity and, together with the Astronomer Royal, Sir Frank Dyson, came up with a plan to test Einstein's theory.

We only see stars at night because during the day their faint light is drowned out by the radiance of the Sun. The stars are still there of course and during a total eclipse, when the Moon blocks the light from the Sun, they become visible. Einstein's theory had predicted that light that passed close to the Sun on its way to Earth would have its path deflected by the warped spacetime around the Sun. This would result in a change in the apparent position of the star that could be measured against the actual position of the star, which was

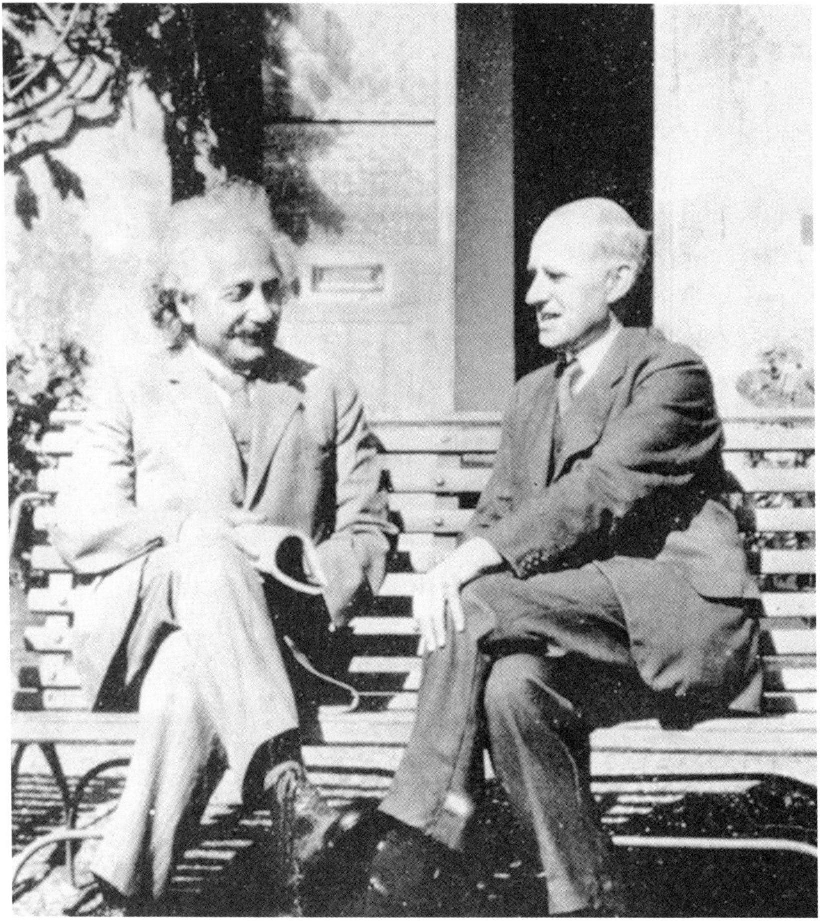

known from observations of the star's position at night. The angle of deflection they were looking for was very small indeed, roughly equivalent to the width of a coin seen from a distance of 3 km, but this was achievable, even with the technology available at the time.

The two expeditions set out from Liverpool in March 1919, one group heading to Brazil and the other to Principe. On the morning of the eclipse it rained heavily on Principe and the required observation seemed hopeless. During the morning the sky began to clear but, as totality (total eclipse) approached, the last few tatters of clouds still threatened to frustrate the observations. Eddington reported that he didn't actually see the eclipse beyond a couple of glances as he was too busy changing the photographic plates in his camera. He was concerned that the cloud would prove to have interfered with the star images.

How did an eclipse prove Einstein was right?

Star appears here during solar eclipse

1.75 seconds of arc

Line of sight

actual light path

True position of the star

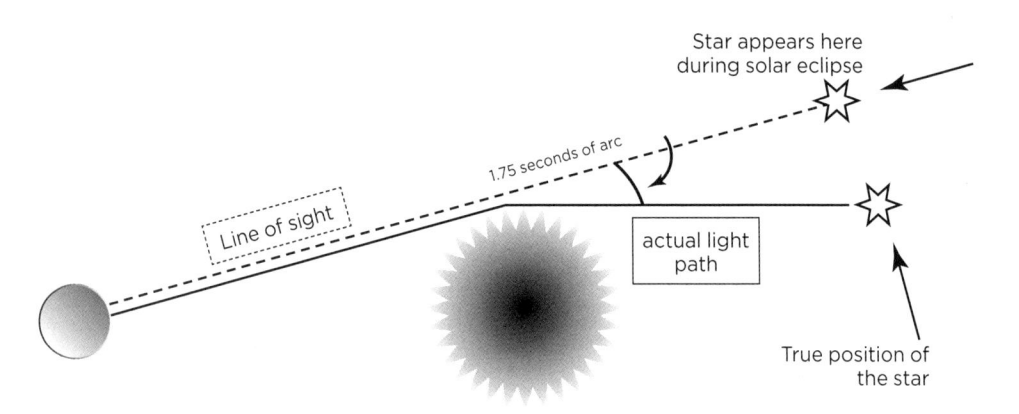

The Sobral team were luckier with the weather. Now the plates from both expeditions would have to be developed and carefully examined. Who was going to be right – Newton or Einstein? One of the Brazil photographs appeared to agree with Einstein, another with Newton. Eddington's plates showed fewer stars, but what was visible seemed to back Einstein. Eddington decided that the Newtonian result from Brazil was due to faulty equipment and declared Einstein vindicated.

Famously, soon after he heard the news, Einstein was with one of his graduate students, Ilse Schneider. She asked him what he would have done if the observations had shown his theory was wrong. Einstein replied, 'Then I would have been sorry for the dear Lord; the theory is correct.'

In other news...

Readers of *The Times* newspaper on 7 November 1919 had some interesting headlines to read. On page 11 they could read about the 'Armistice and treaty terms', 'Devastated France' and 'War crimes against Serbia'. On page 12 there was more news concerning the aftermath of the war, including news of the 'Armistice day observance/Two minutes' pause from work'. Casting an eye on column six, the reader would find some world-changing news of a different kind: 'Revolution in science/New theory of the universe/Newtonian ideas overthrown'. And there, halfway down the column, a subheading that must have resulted in much scratching of heads: 'Space warped'.

Einstein himself wrote an article for *The Times*, published in the issue of 28 November. With odd prescience, given subsequent events in Germany, he concluded it:

'By an application of the theory of relativity to the tastes of readers, today in Germany I am called a German man of science and in

LIGHTS ALL ASKEW IN THE HEAVENS

Men of Science More or Less Agog Over Results of Eclipse Observations.

EINSTEIN THEORY TRIUMPHS

Stars Not Where They Seemed or Were Calculated to be, but Nobody Need Worry.

A BOOK FOR 12 WISE MEN

No More in All the World Could Comprehend It, Said Einstein When His Daring Publishers Accepted It.

England I am represented as a Swiss Jew. If I come to be regarded as a bête noire, the descriptions will be reversed and I shall become a Swiss Jew for the Germans and a German man of science for the English!'

A message from Mercury

Astronomy played another role in demonstrating the validity of

> **WHO'S THE OTHER ONE?**
>
> There is a story, probably apocryphal, concerning Arthur Eddington. At one of his lectures someone suggested that he, Eddington, was one of only three people who understood general relativity. Eddington paused for a moment before responding, 'I'm trying to think who the third person is.'

the general theory. A long-standing puzzle for astronomers was the fact that the orbit of Mercury, the planet closest to the Sun, did not quite fit in with Newton's equations.

As the planets orbit the Sun they follow an elliptical path, as determined by Kepler in 1609 and explained by Newton some 50 years later. The elliptical path means that the planet has a point of closest approach to the Sun (astronomers call this the perihelion). This point doesn't always occur in the same place on each orbit but, because of the pull of the planets on each other, an effect predicted by Newton, the perihelion slowly moves around the Sun. This rotation of the orbit is called a precession.

How did an eclipse prove Einstein was right? | **171**

The problem was that Newton could explain the precession of all of the planets except that of Mercury. Mercury's rate of precession was just a little bit more than Newton predicted. It was a small difference but not one that could be ignored.

On Christmas Eve 1911, Einstein wrote:

'At this time I am [again] busy with considerations on relativity theory in connection with the law of gravitation... I hope to clear up the so-far unexplained changes of the perihelion length of Mercury... [but] so far it does not seem to work.'

Astronomers hunted for a way to explain Mercury's odd behaviour. Perhaps there was a swarm of asteroids between Mercury and the Sun, or perhaps even an undiscovered planet, that was tugging on Mercury as it orbited. There were many ideas but none seemed to answer all the questions. What they all had in common though was that they accepted Newton's law of gravitation as accurate.

In 1916, Einstein stepped in with the equations from his newly forged general theory of relativity. He was able to show that his concept of how gravity worked exactly predicted Mercury's orbital movements. The reason for the discrepancy was the warping of spacetime so close to the huge mass of the Sun. Einstein was jubilant at this evidence that his theory was right and his calculations agreed with the astronomers' observations. 'For a few days, I was beside myself with joyous excitement,' he wrote.

Chapter 15

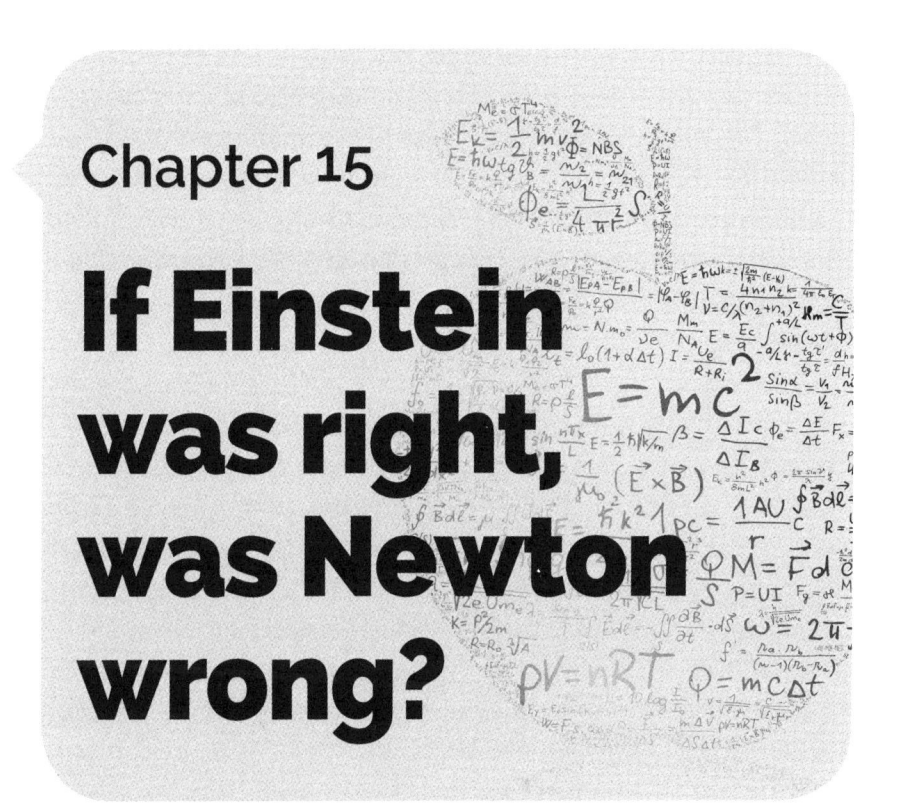

If Einstein was right, was Newton wrong?

The very nature of scientific enquiry is that it never stands still, no truth is absolute, no proof immune from being overturned.

'Newton, forgive me; you found the only way which, in your age, was just about possible for a man of highest thought and creative power. The concepts which you created are guiding our thinking in physics even today, although we now know that they will have to be replaced by others farther removed from the sphere of immediate experience, if we aim at a profounder understanding of relationships'

– Albert Einstein.

Isaac Newton may have been wrong in the tiniest details but for 200 years or so he was undoubtedly right in practice. It is the very nature of scientific enquiry that truths are never absolute but simply the best description of reality that can be achieved with the knowledge available at the time. A law of science is an explanation or a statement that always appears to be true. But science doesn't stand still. Newton's law of gravity and his three laws of motion are excellent at explaining why objects move the way they do. However, in 1905, Albert Einstein came along and showed that, for objects travelling at speeds approaching that of light, Newton's laws didn't apply any more. Newton wasn't wrong – he just couldn't imagine or foresee the limits of his laws.

Newton offered a way of calculating the effects of gravity but he never sought to explain it. In 1687 he wrote:

'I have not yet been able to discover the cause of these properties of gravity from phenomena and I frame no hypotheses [...]. That one body may act upon another at a distance through a vacuum without the mediation of anything else, by and through which their action and force may be conveyed from one another, is to me so great an absurdity that, I believe, no man who has in philosophic matters a competent faculty of thinking could ever fall into it.'

– Isaac Newton

Einstein wanted to find an explanation for the phenomenon of gravity, and the explanation he found just happened to provide much more accurate predictions than Newton did, but Newton's laws still work perfectly well in 'normal' (that is, at well below the speed of light) conditions – they're certainly accurate enough to plot a course to send a probe from the Earth to Pluto. Einstein didn't prove Newton was wrong; he produced a theory of gravity that extended to cover situations Newton had no conception of.

It is one of the hallmarks of science, and one of its guiding principles, that a theory is nothing if it can't stand up to experimental scrutiny. As Richard Feynman said: 'It doesn't matter how beautiful your theory is, it doesn't matter how smart you are, it doesn't matter what your name is. If it disagrees with experiment, it's wrong.'

Of course, it takes a great deal of experiment to prove a theory right or wrong. One experiment isn't enough, results have to be replicated and verified. Newton's laws stood up to experiment time and time again for over a hundred years. For example, 19th-century astronomers noticed that Sirius, the brightest star in the night sky, appeared to wobble slightly in its path. Newton's laws say that if something isn't moving as you expect it to there must be a force being applied. Some might have put forward the alternative argument that this showed that Newton's laws only hold in our solar system and not in interstellar space. However, if Newton was right then perhaps there was an unseen star orbiting Sirius whose gravity was causing it to wobble. In 1862, this star was discovered. In fact, Sirius is a binary star system – two stars orbiting a point between them. One is a main sequence star, Sirius A, and the other a white dwarf star, Sirius B. Hence the wobble – a vindication for the laws of Newton.

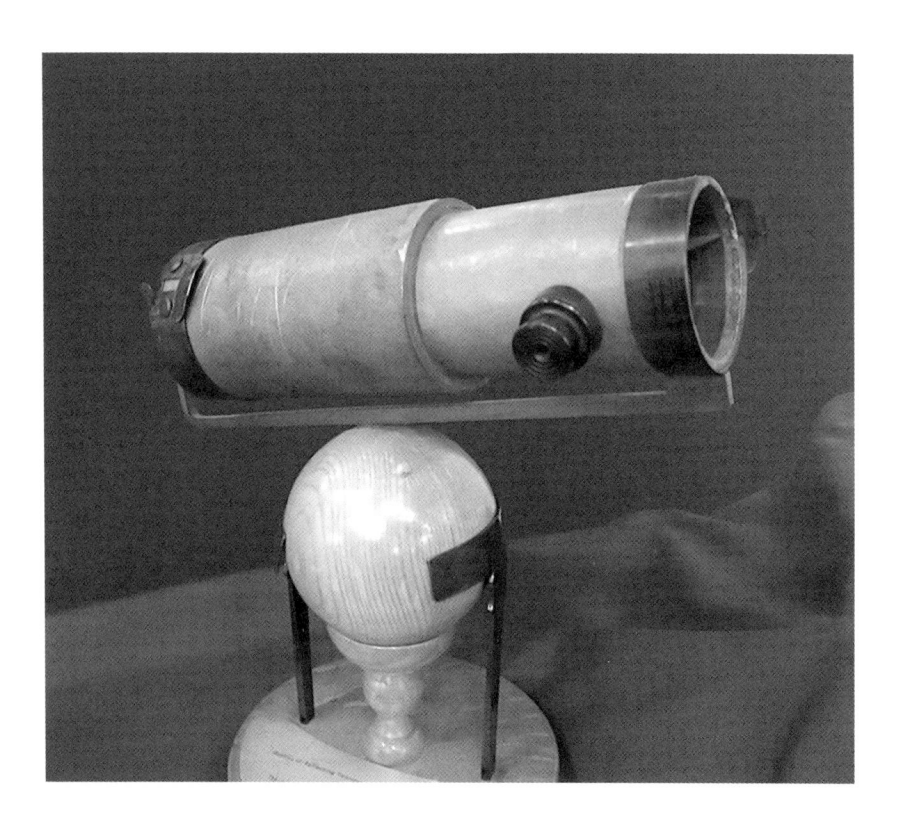

One of the first real cracks in Newtonian physics was the unexplained deviations in the orbit of Mercury. There was nothing in Newton's *Principia* that explained that, no matter how hard people tried. Though astronomers searched for a new planet to balance the gravitational books, even calling it Vulcan, there was no intrasolar version of Sirius's companion star, Sirius B, to save Newton.

The predictions of the general theory of relativity are the same as those of Newton's theory of gravitation so long as the gravitational fields involved are weak, in other words so long as the velocities of all objects interacting with each other gravitationally are small compared with the speed of light. A gravitational field is considered strong if the escape velocity required to break free of it approaches the speed of light. All gravitational fields encountered in the solar system, even the one in the vicinity of the Sun, are weak by this definition. At low speeds and in weak gravitational fields, general and special relativity's predictions agree with everyday experience and Newtonian physics.

Einstein had huge respect for Newton's achievements. He wrote of him: 'In one person he combined the experimenter, the theorist, the mechanic, and, not least, the artist in exposition... He stands before us strong, certain, and alone: his joy in creation and his minute precision are evident in every word and every figure.'

So, yes, Newton was wrong but, so far as anyone could know, he was right in his time; and yes, Einstein is right for now, but one day he too may be shown to be wrong if some yet deeper truth is discovered. As we shall see, even general relativity is not the full story. At the extremes of the cosmos, in places like black holes where the

ESCAPE VELOCITY

The idea of an escape velocity was one that arose from a study of Newton's laws. Ignoring complicating factors such as air resistance, the escape velocity is the speed an object would need to attain in order to escape a planet's gravitational pull and continue on into space. For example, the escape velocity from Earth is about 11.2 km/s at the surface. When it became known that light had a finite speed, physicist John Michell posed an interesting question in 1783. Could we not in theory have a very small, but very massive object, with an escape velocity so high, that light would not be able to escape from it? If this were so, said Michell, then the most massive objects in the Universe might well be dark. This may be the first reference to the phenomenon now known as a 'black hole'.

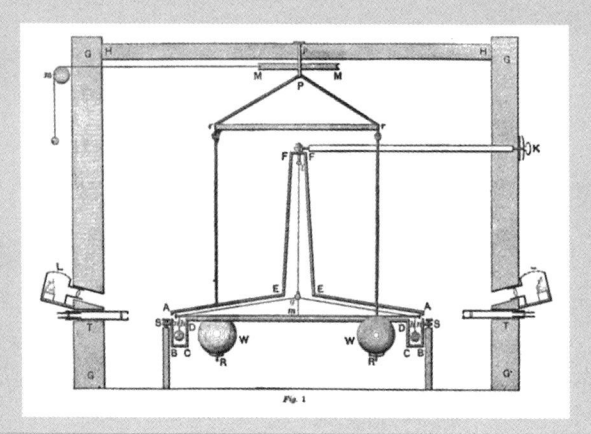

Fig. 1

escape velocity not only approaches but exceeds the speed of light, Einstein begins to fail us. General relativity also breaks down on the subatomic scales where we enter the quantum realm.

Like Newton before him, Einstein may one day be seen as another giant step along the way to a fuller understanding of the workings of the Universe. For now, though, every test that we have tried indicates very strongly that general relativity reflects the way that nature seems to work. Tests of the general theory's equality between gravitational and inertial mass have been found to be accurate to within one part in 10 trillion, which is as accurate as we can be with current equipment. There are no cracks observable in general relativity that we might try to slip a new theory into. General relativity does the job of explaining the Universe. But 200 years ago they said the same about Newton's law of gravity!

As French philosopher Claude Lévi-Strauss put it: 'The scientist is not a person who gives the right answers, he's one who asks the right questions.'

Chapter 16

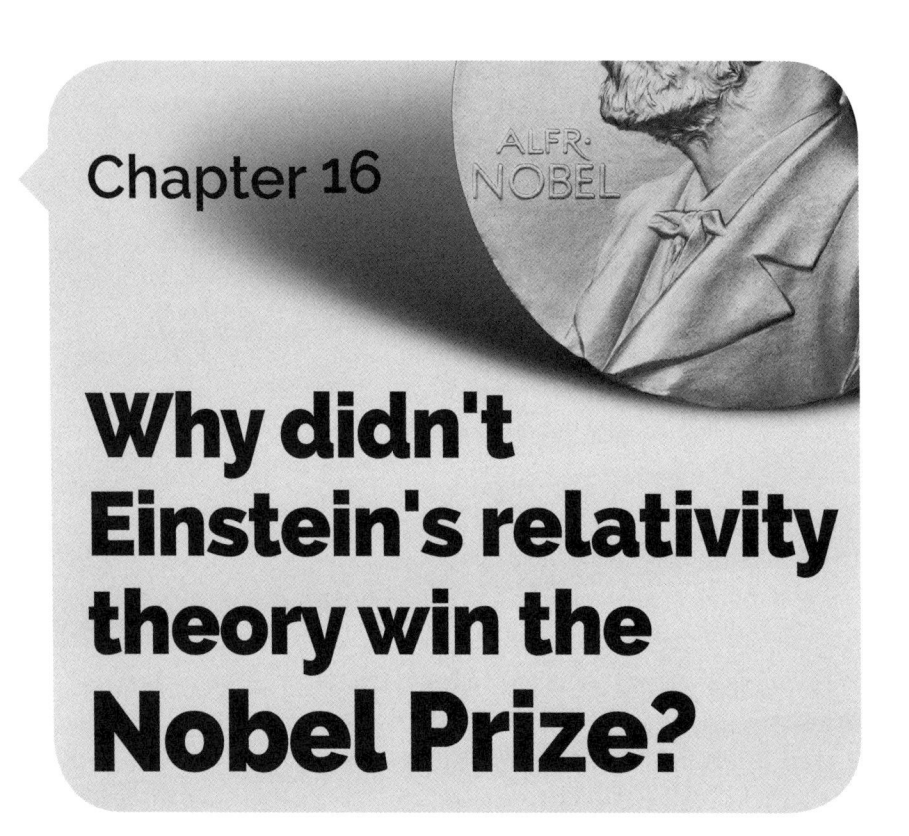

Why didn't Einstein's relativity theory win the Nobel Prize?

Politics of the time – the biases and special interests affecting the Nobel Prize committee.

When Albert Einstein published his ground-breaking papers on the nature of space, light, motion and the atomic realm in 1905 he was a 26-year-old patent clerk, little known outside his immediate circle. In contravention of accepted norms, Einstein published his special relativity paper without referencing other sources. All other scientific papers had cited work by other scientists, but Einstein wasn't really aware of anyone who had been thinking along the same lines as he had. He was always rather reluctant to acknowledge that the steps that others had been taking to explain the results of the Michelson–Morley experiment had any influence on his thinking.

Why didn't Einstein's relativity theory win the Nobel prize?

Einstein was disappointed by the lukewarm reaction to special relativity. The first paper on special relativity, other than by Einstein, was written in 1908 by Max Planck, and it was largely due to the importance of Planck, as opposed to that of a technical expert third class in the Bern patent office, that acceptance of relativity began to grow. Also in 1908, Hermann Minkowski published an important paper on relativity, and showed that the Newtonian theory of gravitation was not consistent with relativity.

Hendrick Lorentz, who had attempted to explain the result of the Michelson–Morley experiment by suggesting that objects contracted as they moved through the ether, never seemed to accept Einstein's conclusions, even though they echoed some of his own thinking. He gave a lecture in 1913 in which he remarked that he found:

'A certain satisfaction in the older interpretation according to which... space and time can be sharply separated... Finally it should be noted that the daring assertion that one can never observe velocities larger than the velocity of light contains a hypothetical restriction of what is accessible to us, a restriction which cannot be accepted without some reservation.'

Nonetheless, Lorentz and Einstein were jointly proposed for the 1912 Nobel Prize for physics for their work on special relativity, the proposer being the German physicist Wilhelm Wein, who had won the prize the previous year for his discovery of the proton.

Why didn't Einstein's relativity theory win the Nobel prize? | **185**

And the award goes to...

The citation for Einstein's Nobel Prize reads: 'The Nobel Prize in Physics 1921 was awarded to Albert Einstein "for his services to Theoretical Physics, and especially for his discovery of the law of the photoelectric effect".' The discovery of the photoelectric effect was indeed important and had profound implications, but weren't special and general relativity even more so?

Nominations for the Nobel Prize are considered by a five-man (and it would have been exclusively men in those days) committee nominated by the Swedish academy of science. They may be received by the prize committee from previous Nobel laureates (anyone can nominate for any field, not just their own), and from professors in select universities, which in the early 20th century meant almost exclusively universities in Nordic- and German-speaking countries. In the decade leading up to 1921, Einstein was nominated repeatedly. Two particular members of the committee were chosen to write reports on his suitability for the prize and repeatedly they recommended that he not be awarded it.

One of these members was the 1903 chemistry prize winner, Svante Arrhenius. Arrhenius was one of the founders of physical chemistry. A physicist himself, Arrhenius was impressed by Einstein's work on Brownian motion and even thought it was worthy of a Nobel Prize. He argued that it would look strange to award the prize for Brownian motion since Einstein's other work surpassed it, but on the other hand he couldn't nominate him for the later work as he regarded that as still experimentally unproven.

Allvar Gullstrand, the winner of the 1911 prize for medicine, was the member of the physics committee who was most vehemently opposed to awarding the Nobel Prize to Einstein. Gullstrand's speciality was the optics of the eye, and his interests were in theoretical optics. Many of Einstein's critics simply failed to understand general relativity, but this wasn't true of Gullstrand; he objected to the theoretical, non-experimental approach Einstein had taken. Like Arrhenius, he argued that there was little in the way of experimental evidence for special relativity. He remarked to a friend at one point that Einstein 'must never receive the Nobel Prize, even if the entire World demands it'.

Gullstrand, like Arrhenius, argued that there was little empirical evidence in favour of special relativity. General relativity had its famous three tests, but one of these, the gravitational redshift of the Sun, was considered unfavourable to Einstein by most experts until 1922. The Report of the Nobel Prize Committee for 1917 refers approvingly to Einstein's work, but also mentions the fact that measurements carried out at the Mount Wilson Observatory in California had not found the redshift that general relativity predicted. 'It appears that Einstein's relativity theory, whatever its merits in other respects may be, does not deserve a Nobel Prize,' the Committee concluded. That the redshift is a real phenomenon was confirmed by laboratory experiments at Harvard University in the 1960s.

Another test of general relativity, the bending of light by the Sun, was widely contested, despite the results of Sir Arthur Eddington's 1919 eclipse expedition. The single greatest demonstrable success of relativity theory, either special or general, was Einstein's explanation of anomalies in the orbit of Mercury that couldn't be explained by Newtonian mechanics. Einstein had shown that his theory predicted a perihelion shift, a change in the point in its orbit at which Mercury came closest to the Sun, which agreed precisely with the observed effect. Gullstrand claimed that Einstein had fudged the calculation to fit the result and that Einstein's theory could match any result one wished for a given problem, which isn't actually true.

Interestingly, while trying to demonstrate the absurdity of relativity, Gullstrand accidentally stumbled upon a very important consequence of it: the viewpoint of an observer falling into a black hole. His conception of a black hole was one in which, once inside the event horizon, space was drawn in towards the singularity faster than light.

Nominations for Einstein for the 1920 prize flooded in after Eddington's results became known, but they were not well received by the committee. According to science historian Robert Friedman, the committee did not want a 'political and intellectual radical, who – it was said – did not conduct experiments, crowned as the pinnacle of physics'. The 1920 prize was given to the Swiss physicist Charles-Edouard Guillaume for his discovery of an inert nickel-steel alloy. When the announcement was made, the previously unknown Guillaume was, again according to Friedman, 'as surprised as the rest of the world'.

Einstein the celebrity

In early 1920, following Eddington's confirmation of gravitational light bending, Einstein became something of a reluctant celebrity. He was always charming and patient in his dealings with people, and he was sought out for his opinions on all manner of things, but he really was more comfortable being left to pursue his work. Few understood what relativity was all about but everyone, it seemed, wanted to talk about it. A great many words were expended trying to explain the principles of general relativity to a lay audience with varying degrees of success; this is something which, a hundred years on, you may judge to be still true.

In the 1920s, the journalist Alexander Moszkowski, a German-Jewish satirist, published a book of conversations with Einstein in which he commented on the public passion for relativity: 'In all nooks and corners, social evenings of instruction sprang up, and wandering universities appeared with errant professors that led people out of the three-dimensional misery of daily life into the more hospitable Elysian fields of four-dimensionality.'

Max Born was horrified that Einstein had agreed to collaborate on the book, fearing it would stoke up the anti-Semitic sentiment against Einstein that was already making itself felt. A growing number of German nationalist scientists had taken to referring to Einstein's ideas as 'Jewish physics'. Einstein himself maintained an air of detachment. 'The whole affair is a matter of indifference to

> **POWER IN THE ATOM**
>
> In 1920, following a lecture Einstein gave in Prague, a reception was given for him by the university physics department. Following a number of enthusiastic speeches, Einstein was invited to respond. Instead of giving the expected speech, Einstein instead announced: 'It will perhaps be pleasanter and more understandable if instead of making a speech I play a piece for you on the violin.' He then went on to perform a sonata by Mozart in what his friend Philipp Frank described as a 'moving manner'.
>
> The next day, according to Frank, a young man approached Einstein in Frank's office. On the basis of $E = mc^2$, insisted the man, it would prove possible 'to use the energy contained within the atom for the production of frightening explosives'. Einstein brushed him off, declaring the idea foolish.

me,' he said, 'as is all the commotion, and opinion of each and every human being. I will live through all that is in store for me like an unconcerned spectator.'

Winning the prize

Although Einstein was again nominated in 1921, Gullstrand blocked that too and convinced the Nobel Committee for Physics that none of the year's nominations met the criteria as outlined in the will of Alfred Nobel. According to the Nobel Foundation's statutes, the Nobel Prize can, in such cases, be held back until the following year, and this is what happened.

The 1921 deferral meant that two prizes could be awarded in 1922. As had been the case in the previous two years, Einstein received many nominations for relativity, but this year there was also a nomination for his work on the photoelectric effect. The nomination came from Carl Wilhelm Oseen, a Swedish theoretical physicist. Oseen wanted the committee to recognize the photoelectric effect as a fundamental law of nature, and not just as a theory. He did so, not so much to support Einstein, as to champion the work of Niels Bohr. Bohr had proposed a new quantum theory of the atom that, according to Oseen, was 'the most beautiful of all the beautiful' ideas in recent theoretical physics. In his report to the committee, Oseen, exaggerating the close ties between Einstein's photoelectric effect and Bohr's new description of the atom in his report to the committee, succeeded in his aim. The committee was won over and, on 10 November 1922, they gave the 1922 prize to Bohr and the delayed 1921 prize to Einstein.

Einstein, en route to Japan when he heard the news, did not attend the official ceremony and didn't actually receive his prize until the following year. Einstein had promised that the monetary award from the Nobel Prize would be put in trust for his sons, with his ex-wife, Mileva Maric, allowed to draw from the interest, and so the prize money was duly transferred over to Maric.

Chapter 17

What was Einstein's greatest blunder?

Einstein introduced a new factor into relativity to make the theory fit the universe as he believed it to be, but was that really his greatest blunder?

It is worth reminding ourselves just how big the Universe was thought to be when Einstein published his general theory of relativity. At that time, most people believed that the Milky Way galaxy was the whole Universe and there was nothing beyond it. The evidence was only just beginning to accumulate that the Universe was much, much bigger than anyone had previously imagined and debates began about whether or not some objects in space might lie outside the Milky Way.

In 1923, Edwin Hubble settled the argument when he used the most powerful telescope in existence at the time, the Hooker Telescope at the Mount Wilson Observatory in California, to make out stars

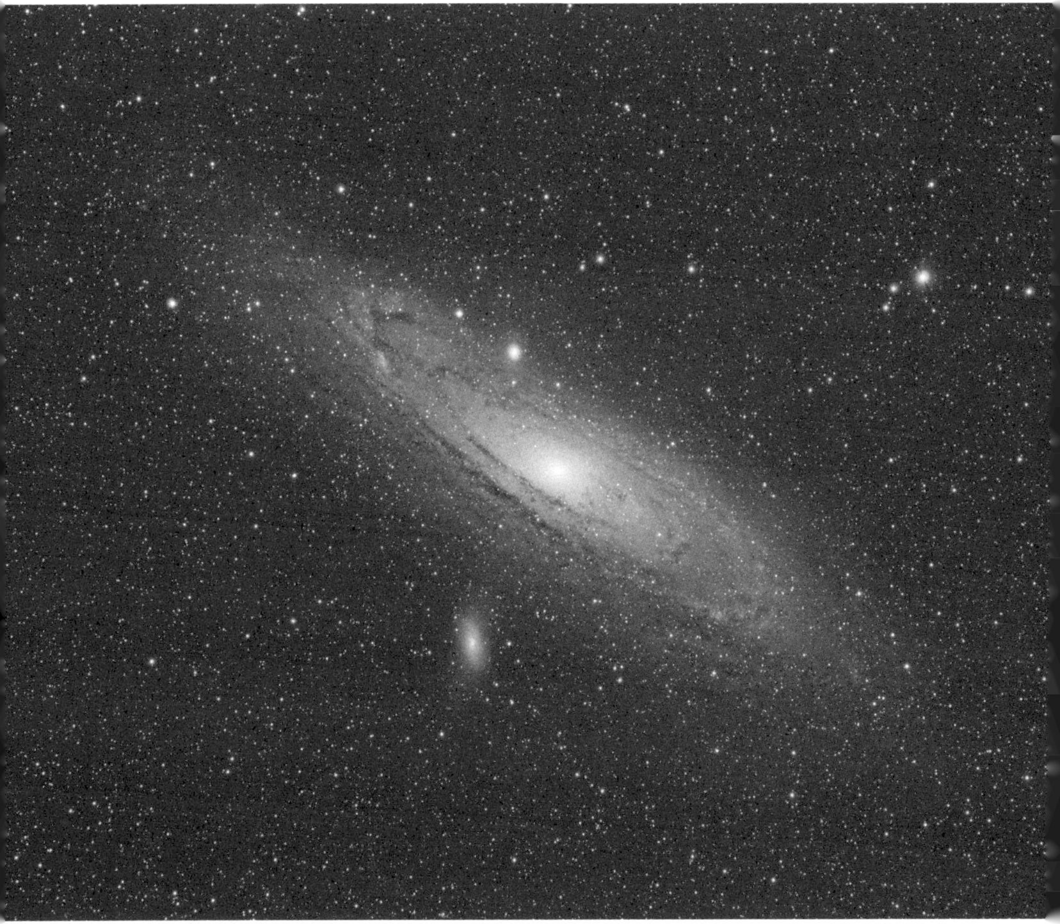

in the Andromeda nebula. He estimated their distance as 800,000 light years (an underestimate of some 1.2 million, as it turned out). Andromeda was a galaxy in its own right, distinct from our own Milky Way galaxy. Hubble went on to find other galaxies that were even more distant. A picture began to emerge of a Universe vast beyond imagining, stretching over billions of light years, with a hundred billion galaxies each containing around a hundred billion stars. We are indeed a very long way from being the centre of creation.

However big it turned out to be, another belief about the Universe still held – that it was static, no one thought that it might expand or contract. In 1929, Hubble became famous overnight when he made another game-changing

> **DOPPLER REDSHIFT**
>
> The redshift Hubble saw in the distant galaxies wasn't a gravitational redshift. This was a different type of shift caused by the Doppler effect. You'll have heard it with sound waves. If a police car shoots past you, siren wailing, the pitch of the siren gets lower as it travels further away from you. This is because successive sound waves take longer to reach you, which sounds like a lowering in pitch. When it is approaching you the opposite happens – successive sound waves reach you faster and the pitch goes up. A similar effect applies to light being emitted by moving objects, but instead of a pitch change the wavelength of the light is shifted towards the red end of the spectrum (redshift) if the object is moving away and towards the blue end (blueshift) if it is moving towards you.

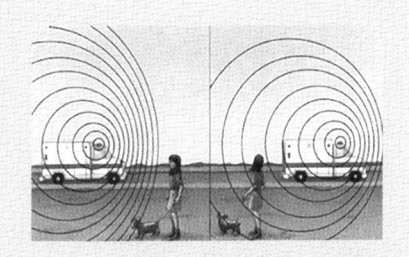

discovery. The light coming to us from the distant galaxies is shifted towards the red end of the electromagnetic spectrum (redshifted, see box), indicating that these galaxies are moving away from our solar system. The further they are away from us, the faster they are retreating. Those that are twice as far move away roughly twice as fast. The best explanation for this was that the Universe is expanding.

The cosmological constant

Einstein's general theory certainly allowed for the notion that the Universe could be either expanding or contracting. In fact, applied to the Universe as a whole, and not just to a star or planet within the Universe, the equations of general relativity demand that the size of the Universe is changing. General relativity doesn't allow for a static Universe – it could not exist as the curving of spacetime by the matter in it would eventually cause the Universe to collapse in on itself. If the Universe was neither static nor collapsing, then it had to be expanding. But Einstein, in common with everyone else at the time, thought this far-fetched.

His concern was that if the Universe were expanding then, logically, it had to be expanding from somewhere. At some distant point in

> **COSMOLOGICAL REDSHIFT**
>
> Doppler shift depends on the motion of the object as it emits energy. Cosmological redshift is a little different. The wavelength at which the light was originally emitted is stretched as it travels through expanding space. Cosmological redshift results from the expansion of space itself and not from the motion of the object that produced the light. The longer the journey the light makes through the expanding Universe the more it will be stretched and the greater the redshift will be.

time, the Universe must have started out as a single point containing all space and time. Einstein thought this was a nonsensical idea and in 1917 he introduced a term called the 'cosmological constant' (he referred to it as a 'slight modification') into his equations. This was a repulsive force that counterbalanced the attraction of gravity and prevented the expansion or contraction of the Universe. Einstein was not entirely happy with this addition, admitting that it was 'not justified by our actual knowledge of gravitation'.

Edwin Hubble's discovery of the redshifting of distant galaxies could not be disputed, however. He had demonstrated that the Universe really was expanding and this was reported in the popular press as a challenge to Einstein's theories. Einstein was happy to admit to his mistake and remove the cosmological constant from his equations, glad that it wasn't necessary after all. He described the astronomers at Mount Wilson as 'outstanding', writing to his friend Michele Besso that the situation was 'truly exciting'. It's a pity that Einstein hadn't trusted his original equations. Had he done so he would have predicted the expanding Universe a decade before Hubble confirmed it.

Arthur Eddington and others pointed out that the cosmological constant would not have worked in any event as it required the Universe to be in such a delicate state of balance, somewhat like balancing a pencil on its point, that the slightest disturbance would have triggered a runaway expansion or contraction.

Why doesn't the cosmos collapse?

Some of the problems faced by the astronomers and physicists of the early 20th century had been discussed over 200 years earlier. In 1692, Newton received a letter from the Reverend Richard Bentley.

Supposing the Universe were infinite, said Bentley, as many supposed it to be, then, that being the case, every part of the Universe should feel the pull of gravity, and surely, therefore, it should collapse in on itself?

Newton tried to explain this conundrum by arguing that if the stars were evenly distributed in space then the force of gravity would act equally in all directions and a balance would be maintained. He quickly realized that this wouldn't do – the slightest movement of any star would upset the balance and the whole cosmic edifice would come crashing down.

Newton and Bentley had made one major error – the stars are not stationary. (It was partly the notion of the 'fixed stars' that allowed Newton to come up with his ideas about absolute space.) It was Edmund Halley, of comet fame and vindicator of Newton's laws, who first observed that a few stars had shifted from the positions recorded for them on Greek star maps.

Olbers' paradox – why isn't the sky full of stars?

Halley pointed out another problem posed by an infinite Universe. If the

Universe were infinite then wherever you looked there ought to be a star – the entire sky should be shining as brightly as the Sun! Obviously, it wasn't – an observation that had led Kepler to conclude that the Universe couldn't be infinitely large. The problem became known as Olbers' paradox, after the German astronomer Heinrich Olbers (1758–1840). He suggested that there must be clouds of dust between the stars, hiding some of them from our view. But this solution was flawed also. Given long enough, the energy from the distant stars would heat up the gas clouds until they glowed and the sky would be filled with light. The answer to the problem came with Edwin Hubble's discovery that the Universe is expanding. The light from the furthest reaches of the Universe has not had time to reach us – and perhaps never will. The Universe is dark because it started with a Bang.

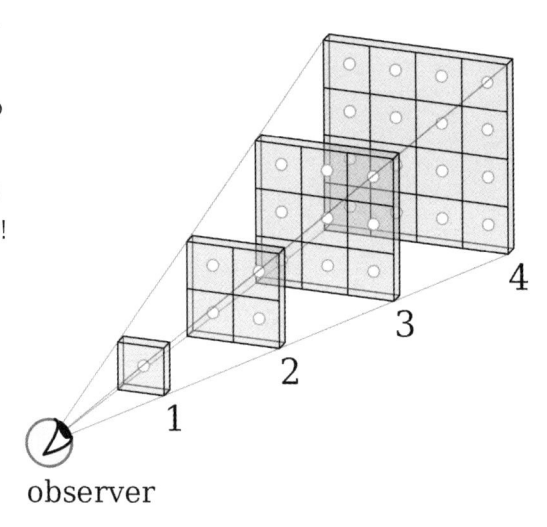

His biggest blunder?

It is an often-repeated tale that Einstein called the cosmological constant his 'biggest blunder'. But did he actually say that? It was the physicist George Gamow who reported that Einstein had used the phrase but there's no hard evidence that he ever did – it certainly never appeared in any of his writings.

There was one mistake that Einstein really regretted and we'll come to that soon. After visiting Einstein in Princeton on 16 November 1954, Linus Pauling wrote in his diary: 'He said that he had made one great mistake – when he signed the letter to President Roosevelt recommending that atom bombs be made.'

The return of the cosmological constant

In the late 1990s, cosmologists made a startling discovery –

the Universe isn't just expanding, it's expanding at an increasingly rapid rate. The cause of this accelerating expansion is a mystery – scientists refer to a 'dark energy' at work. Most observations support the idea that this dark energy behaves like Einstein's 'cosmological constant' and many cosmologists are keen to revive the term. One speculation is that pairs of so-called 'virtual' particles and antiparticles pop in and out of existence in empty space, a phenomenon allowed for by quantum mechanics. The energy carried by these particles might exert a repulsive force that pushes everything in the Universe outwards. The cosmological constant, far from being a blunder, might cause scientists to have to re-evaluate what they believe to be true about cosmology, particle physics and the fundamental forces of nature.

> **EINSTEIN'S ENVELOPE**
>
> On his second trip to the United States, in 1931, Einstein visited the Mount Wilson Observatory with Edwin Hubble. They were met there by the now frail and elderly Albert Michelson, of the Michelson–Morley ether experiment. Einstein's wife, Elsa, accompanied him and, as the telescope was being demonstrated, she was informed that it had been used to determine the size and shape of the cosmos. 'Well,' she responded, 'my husband does that on the back of an old envelope.'

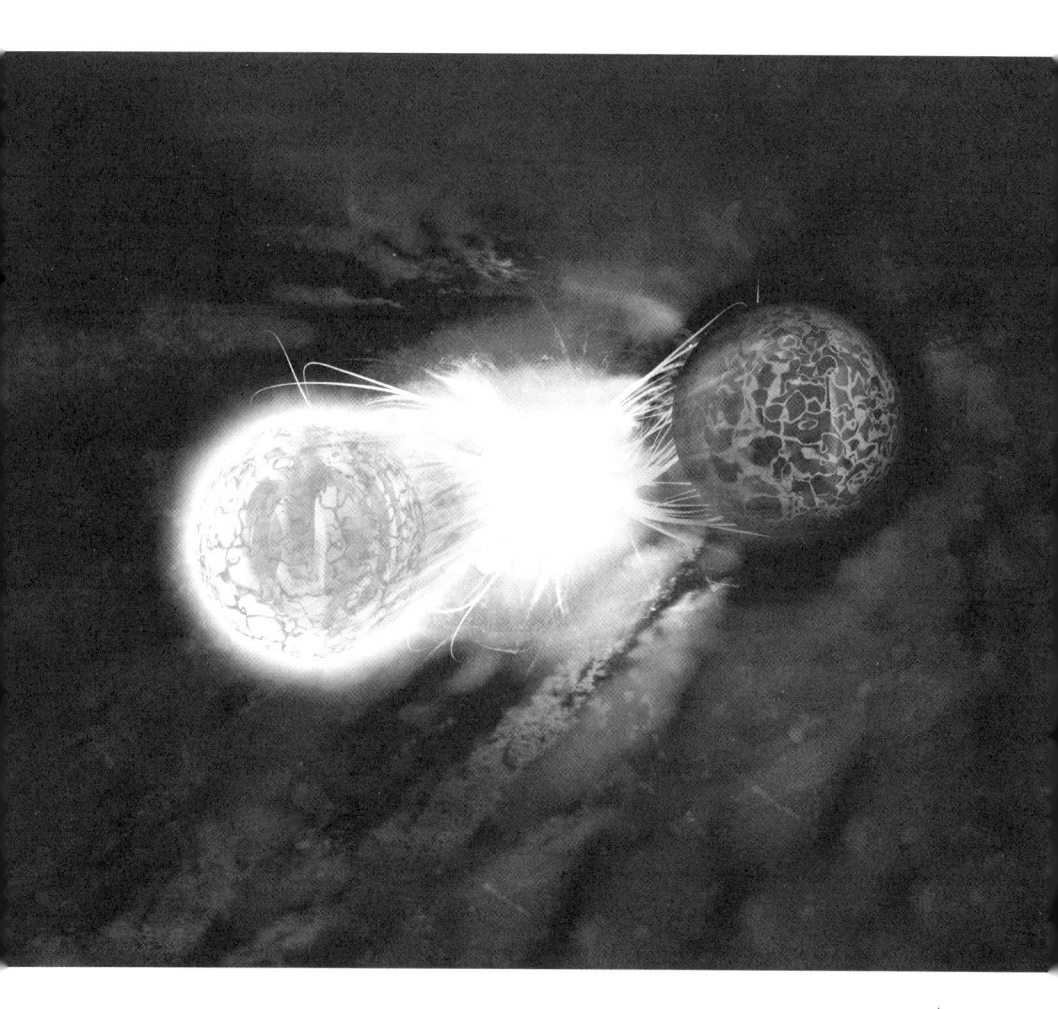

What was Einstein's greatest blunder? | **207**

Chapter 18

Where does Einstein's relativity theory break down?

What happens to relativity at the most extreme limits of reality, inside the event horizon of a black hole?

General relativity has been an incredibly successful means of providing new insights into our understanding of the Universe and the way it works. But just like Newton before him, there are things that Einstein cannot explain.

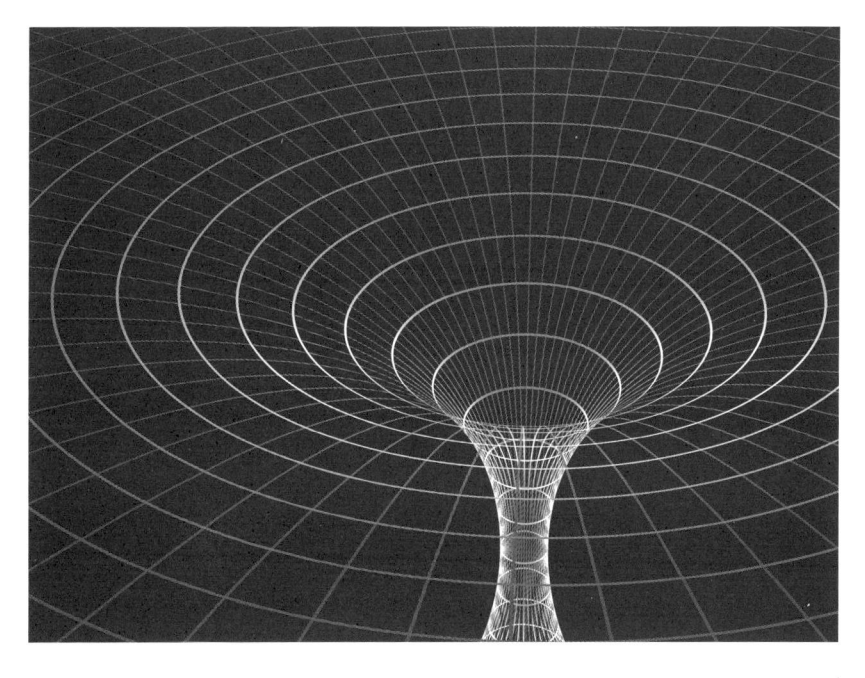

The more compact and massive the object, the stronger its gravitational influence. General relativity predicts the existence of black holes, regions where the density of matter is so intense that spacetime is warped and curved to the extent that it becomes infinite. So deep is the gravity well formed by a black hole that nothing can break free from it. It is like a hole in spacetime and not even light has a spacetime path it can follow to escape it.

Pulsars and neutron stars

In 1967, Jocelyn Bell, a graduate student at Cambridge University, noticed something unusual in her radio telescope observations. There seemed to be a rapidly pulsing signal coming from a point in the sky. She looked further and found another, regularly pulsing, signal. Some people suggested that these might be a sign of intelligent life and they were dubbed LGMs (little green men).

American astronomer Thomas Gold wondered whether they were actually neutron stars. Astronomers had speculated for some time about the existence of neutron stars. Once their nuclear fuel is exhausted, stars that are between five and forty times as massive as our Sun end their lives in a gigantic outpouring of matter and energy called a supernova, as the outer layers of the star are blasted out into space, leading to such a dramatic increase in the star's brightness that it might outshine the other stars in its galaxy for a short time.

At the same time as the outer layers are expanding into space, the star's core collapses. Within seconds, the core's density increases so enormously that the electrons and protons that make it up are squeezed together to form neutrons and the core region becomes an unbelievably dense ball of nuclear matter, where trillions of tons of material are squeezed into each cubic centimetre. A neutron star is no more than 20 km in diameter, but with a mass greater than that of our Sun. If you could bring a teaspoonful of neutron star material to Earth it would weigh more than a mountain.

As the star collapses it spins faster and faster, like a figure skater on the ice spinning faster by pulling her arms and legs closer to her body. The contracting neutron star may eventually be rotating a few hundred times every second. The star's magnetic field becomes more concentrated and more powerful. Electrons within the magnetic field are accelerated to close to the speed of light and emit beams of electromagnetic radiation from the star's north and south poles. The star is now acting like a cosmic lighthouse, with two narrow beams of electromagnetic waves pointing in opposite directions. We can only detect the neutron star if one of these beams happens to sweep across the surface of the Earth and be detected by radio telescopes. Because such stars appear to pulse as the beam briefly points at the Earth, they were given the name pulsars.

Neutron stars, and pulsars in particular, are ideal cosmic

Where does Einstein's relativity theory break down? | **211** laboratories for testing general relativity. The gravity of a compact, massive body such as a neutron star is very strong and therefore the effects of general relativity become much more apparent. For instance, binary systems involving a neutron star orbiting around an ordinary star can be used to make precise measurements of the influence of gravity on light.

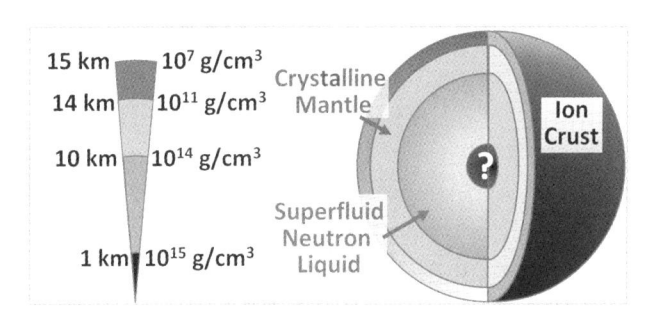

Diagram labels:
- 15 km — 10^7 g/cm³ — Crystalline Mantle
- 14 km — 10^{11} g/cm³
- 10 km — 10^{14} g/cm³ — Superfluid Neutron Liquid
- 1 km — 10^{15} g/cm³
- Ion Crust

Black holes

What would happen if a neutron star continued to collapse? In 1928, the Indian astronomer Subrahmanyan Chandrasekhar calculated that, if a star was bigger than a certain size, the power of its gravitational force would be greater than its atomic particles could resist. The star would simply go on collapsing until it formed a single point, a singularity, warping spacetime around it to such an extent that nothing could escape, not even light. It would become a black hole in space.

A black hole isn't a physical object, it is more a region of spacetime with very peculiar properties. The border that separates this region from the rest of the Universe is called the event horizon. The event horizon is a one-way door – matter or energy can cross over it from the outside, but never comes out again. Therefore an observer will detect no light from a black hole. This means it is impossible to observe one directly but it is possible to see the effect a black hole has on its surroundings.

The light of stars and galaxies passing by a black hole is bent by the black hole's gravitational influence, much as the Sun's gravity bends the starlight that passes it, but to a much greater degree. If an object and a black hole are precisely aligned, an observer would see the light from the object bent to form a ring around the black hole. This phenomenon is called an Einstein ring. If the star is out of alignment – even if only slightly – the ring can't be seen and therefore the black hole is much more difficult to detect.

Astronomers need to look for indirect ways to deduce the presence of a black hole, for example by looking for unexpected movements in nearby stars. If a black hole forms from a star that was part of a binary system it may start to pull gases in towards itself from the outer layers of its neighbouring star. This gas swirls in around the black hole forming an accretion disc that reaches such high temperatures that it emits X-rays. A neutron star can have a similar effect but sometimes observations suggest that the object is too compact to be a neutron star. In that event, astronomers think they

WHAT IS A SINGULARITY?

One of the consequence of Einstein's description of gravity in terms of curves in spacetime is that it allows for the formation of singularities. A singularity is a point where some property is infinite. For example, the density of the material at the centre of a black hole is infinite, because the mass of the star has been compressed into zero volume under the effect of infinite gravity. At the centre of a black hole, spacetime has infinite curvature and space and time cease to exist in any meaningful sense. The laws of physics as we know them break down in the singularity, including relativity. It has to be said that singularities were anathema to Einstein, who believed that there was no place for such infinities in a proper mathematical description of the Universe. He argued that singularities could not appear in nature 'for the reason that matter cannot be concentrated arbitrarily ... because otherwise the constituting particles would reach the velocity of light.'

have a good case for saying they've found a black hole. To the best of our knowledge, the nearest black hole is more than a thousand light years distant, a thought that might help you sleep a little better.

How big is a black hole?

A black hole can be any size. Black holes that form when a large star becomes a supernova have a radius of around 5 km. Galactic black holes, which form in the cores of some galaxies, can have a mass equal to that of millions of stars and be bigger than our solar system. At the other end of the scale, mini black holes, which formed in the early Universe, may be smaller than a grain of sand but be as massive as mountains. The outer boundary of a black hole – its event horizon – forms at a point called the Schwarzschild radius. This is the radius below which the gravitational attraction between the particles making up an object become so intense that it will suffer an irreversible gravitational collapse, eventually forming a black hole. Theoretically anything can form a black hole if you squash it down hard enough. The Schwarzschild radius for an average human being is about 10–23 cm – smaller than the nucleus of an atom.

The Schwarzschild radius was discovered by German astronomer Karl Schwarzschild in 1916 while he was studying Einstein's general relativity equations. It was within months of Einstein publishing his theory that Schwarzschild used it to describe how spacetime would

SUPERMASSIVE BLACK HOLES

If you thought the idea of an ordinary black hole was staggering enough then you ain't seen nothing yet (well you wouldn't – it's a black hole...). Not long after the invention of the radio telescope, astronomers were finding evidence of highly energetic radio galaxies. Like a neutron star, these shoot out beams of highly energetic particles in opposite directions, but on a vastly greater scale. When these beams interact with clouds of intergalactic gas they cause them to emit radio waves which can be detected by telescopes on Earth. It was clear that there was only one possible power source: matter falling towards a compact mass and forming a high-energy accretion disc. But it was on a scale so enormous that the central mass would have to be extremely massive and extremely compact. Astronomers now believe that these are supermassive black holes with more than a million times the mass of the Sun and are found in the core of galaxies, even relatively peaceful ones such as our own Milky Way. The current supermassive heavyweight champion weighs in at a mass of 21 billion Suns, which is pretty much incomprehensible on any human scale of understanding. It can be found in the crowded Coma galaxy cluster, which consists of more than 1,000 galaxies.

be warped near a spherical star. At the time, Schwarzschild was unable to present his findings to the Prussian Academy because he was busy calculating trajectories for the German army's artillery on the Russian front, so he sent his work to Einstein, who presented it on his behalf.

Because of the severe warping of spacetime, strange effects occur at the event horizon of a black hole. An observer watching someone fall towards the event horizon – if such a thing were possible – would see their clocks running slower and slower until, at the event horizon itself, time appeared to freeze. For the person falling, the opposite would be true: they would see time in the rest of the Universe speed up and perhaps even witness its end before crossing the event horizon.

Einstein himself did not believe that black holes would form but other theorists showed how a sufficiently massive star would inevitably collapse at the end of its life to form a super-massive singularity where all the laws of physics, including Einstein's, would break down.

Chapter 19

How did relativity lead to a Big Bang?

Relativity suggested that the universe hadn't existed forever, it had a beginning.

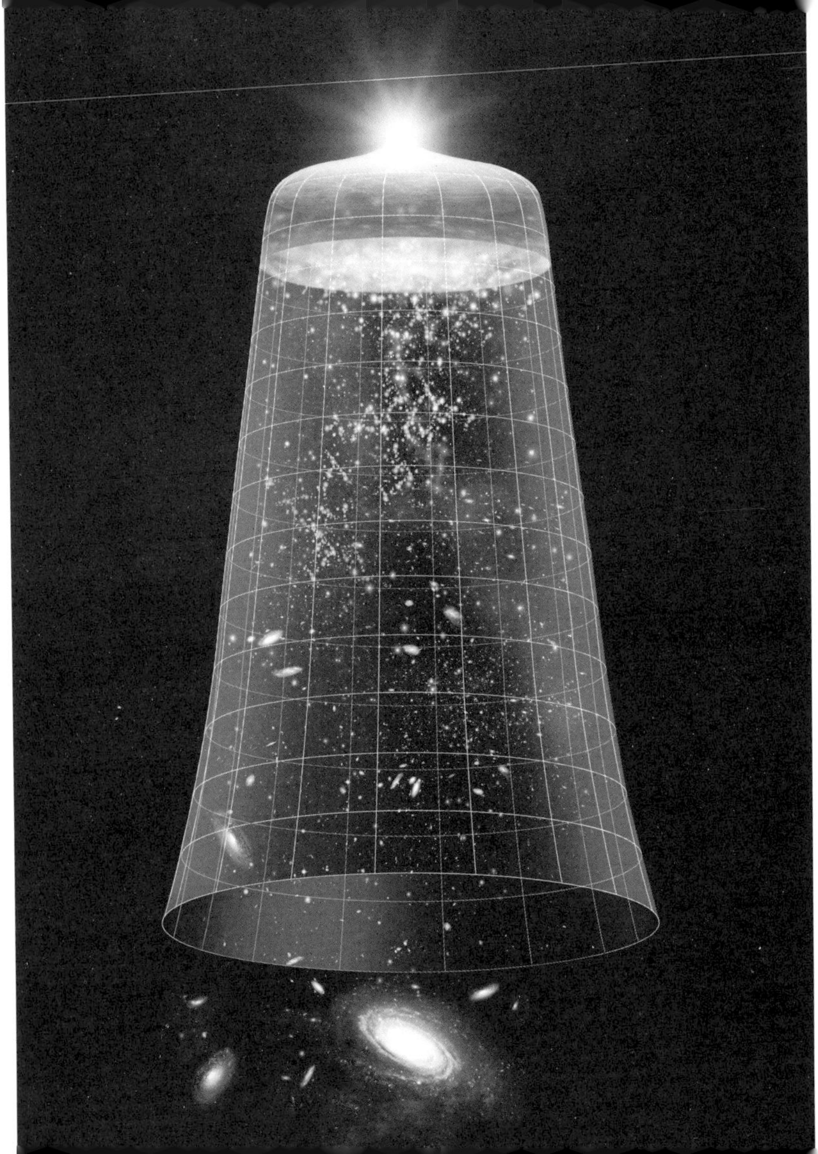

After Einstein introduced the world to general relativity, a number of scientists, Einstein included, tried to see how the theory would apply to the Universe as a whole. At the time this required that they make an assumption about how matter in the Universe was distributed: firstly, if you look on a large enough scale, that the Universe looks more or less the same in whichever direction you look and, secondly, that the Universe looks the same wherever you are in it. That is, the matter in the Universe is homogeneous (the same everywhere) and isotropic (the same in every direction) when averaged over very large scales. This is called the Cosmological Principle.

Armed with Einstein's description of how gravity works and an idea of how the matter is distributed in the Universe, we can start to build a picture of how the Universe has evolved over time. The picture we get is one in which the Universe started out from effectively zero. If we take the expanding Universe and throw it into reverse, rewinding its evolution back in time, we see all matter, all energy, all space and all time contract down into a single point of infinite density and gravity and zero size – a singularity, in other words. At this point, Einstein's equations, the foundation stones of his general theory describing how the distortions of spacetime affect the matter and energy embedded in it, break down – just as they do in a black hole singularity. For whatever reason, we just don't know why, everything that makes up the Universe today expanded out from this zero point in an event that came to be known as the Big Bang.

How did it all begin?

The first to suggest a possible beginning for the expanding Universe was a Belgian priest and astronomer called Georges Henri Lemaître (1894–1966). He put forward his idea of a 'primordial atom' at the end of the 1920s in a famous paper entitled, 'A Homogeneous Universe of Constant Mass and Increasing Radius accounting for the Radial Velocity of Extra-Galactic Nebulae'. Lemaître started with a solution to Einstein's equations corresponding to an expanding Universe. From this, Lemaître deduced the fact that the speed of the far galaxies is proportional to their distance – a finding that Hubble's redshift calculations (see page 199) were in agreement with. Lemaître suggested that in the distant past all the mass in the Universe had been concentrated into a single superatom. According to Lemaître, this primordial atom began dividing again and again, eventually giving rise to all the matter we see today. He didn't use the expression 'Big Bang', but he did talk about a 'day without a yesterday'.

Lemaître met Einstein for the first time in October 1927, during the Fifth Solvay Congress of Physics in Brussels. Einstein had read Lemaître's paper and they discussed it together. Einstein had no criticism regarding the mathematics involved – Lemaître's work was flawless – but he disagreed with the way he had interpreted it, going so far as to call it 'abominable'. Unfortunately for Lemaître, this was still a time when Einstein clung to the idea of a static Universe and his cosmological constant.

When they met again in 1933, Einstein was more receptive, having now abandoned the cosmological constant. Einstein had accepted the idea of an expanding Universe but not that of an initial singularity. He suggested that Lemaître modify his model, hoping that if he made certain changes to it the initial singularity could be avoided. Lemaître soon showed that the revised model also led to a singularity.

A cooling Universe

Following the Second World War, Ralph Alpher and George Gamow suggested that, in the beginning, the Universe was formed from a hot soup of atomic particles at a temperature of trillions of degrees. They called this atomic soup 'ylem'. This cooled as it expanded as its energy was spread over greater and greater volumes of space. Alpher and Gamow, who was his PhD supervisor, wanted to prove that the initial conditions they imagined could explain the various abundances of the elements in the Universe. Some elements, such as hydrogen and helium, are very common, whereas others, such as tin and gold, are much rarer.

They spent several months on their calculations, working out the effects of the fall in temperature and density as the Universe expanded. Their results suggested that hydrogen and helium should be by far the most common elements and that there should be ten atoms of hydrogen for every one atom of helium. This was exactly the ratio that had been determined by astronomers.

In 1948, Alpher, this time with Robert Herman, published a paper in which they predicted that radiation from the early beginnings of the Universe should still be detectable. They calculated that this 'cosmic background radiation', as it was called, would have a temperature of about minus 268°C. This was the last faint glow left over from the unimaginable burst of energy that had given birth to the Universe. Alpher tried to persuade astronomers to search for these echoes from the beginning of time. Unfortunately, the equipment didn't then exist to refute or verify their theory and so the prediction was more or less forgotten for nearly 20 years.

In 1964, US radio astronomers Arno Penzias and Robert Wilson made the discovery that finally tipped the scales in favour of the Big Bang. While testing an astronomical microwave detector called the Holmdel Horn Antenna, they were concerned to discover that the device seemed to be picking up noise from all over the sky. At first, they thought that pigeon droppings might be causing it to malfunction! But after cleaning out the detector – and having the pigeons shot! – they found that, whatever it was, the noise was coming from outside the atmosphere, and from every direction. It never varied, whatever time of day they tried.

WHAT'S IN A NAME?

It didn't help Alpher's case that at the time a great many astronomers still refused to accept that the Universe actually had a beginning. Not long afterwards, in 1950, the British astronomer Fred Hoyle gave a radio talk on the subject. Hoyle was a fierce opponent of the expanding Universe theory, preferring his own theory, which he called 'Steady State', in which the Universe remains much as it always has been due to the continuous creation of matter. He mocked the ideas of people like Alpher and Gamow, referring to their theory as a 'big bang'. The term lodged in people's imaginations and from then on, the idea that the Universe started from an initial point was called the 'Big Bang' theory.

Evolutionary Theory:
Density of matter decreases over time

Steady State Theory:
Density of matter is constant over time

Also around this time, physicists Bob Dicke and Jim Peebles were planning an experiment to test the ideas of Ralph Alpher that there should still be traces of radiation from the early Universe. They expected that light from the earliest stars would have been so massively redshifted on its epic journey across the Universe that it would appear to us as microwave radiation. Penzias and Wilson, hearing of their work, got in touch to ask if what they had found might just be what Dicke and Peebles were looking for. Dicke confirmed that the mysterious signals were indeed the cosmic background radiation and therefore proof of the Big Bang. 'We've been scooped', Dicke admitted.

'When we first heard that inexplicable hum, we didn't understand its significance, and we never dreamed it would be connected to the origins of the Universe,' said Penzias. 'It wasn't until we exhausted every possible explanation for the sound's origin that we realized we had stumbled upon something big.'

Where do we go from here?

Since we've established that we are in an expanding Universe, what happens next? The fate of the Universe depends on the balance

> ### IT WASN'T REALLY A 'BANG'
>
> The Big Bang was not a sudden explosion of all the matter in the Universe out into space. Before the Big Bang there was no space for anything to explode into. Space, time and everything else came into existence with the Big Bang. There is no 'centre' from which everything expanded and with the best time machine your imagination can conjure you still couldn't go back and watch the Big Bang happen. There is no vantage point from which you could see it. The Big Bang was an eruption of space and time that carried all the mass and energy of the Universe along with it. The Universe, by definition, comprises all of space and time as we know it, so it is meaningless and ultimately unanswerable to speculate about what the Universe is expanding into.

between the rate of expansion, which is expressed by a factor called the Hubble Constant, and the curving of spacetime by gravity, which is determined by the amount of matter in the Universe. There are three possibilities. First, the amount of matter in the Universe is greater than the 'critical density' as cosmologists call it, and the expansion is slowed down, stopped and reversed by gravity. Spacetime bends in on itself like a four-dimensional cosmic beachball and the Universe eventually collapses again in a Big Crunch. Second, the density of the Universe is a little less than the critical density.

The Universe continues to expand but at a slower and slower rate. Third, and this appears to be what actually is happening, the rate of expansion gets faster. The gulfs between galaxies, already huge beyond imagining, get ever wider.

Results from the WMAP spaceprobe and observations of distant supernovae seem to indicate that the expansion of the Universe is actually accelerating, which implies the existence of an unknown force acting counter to gravity, sometimes referred to as 'dark energy'. The existence of this force is very reminiscent of Einstein's cosmological constant.

The amount of matter in the Universe also determines its geometry. If the density of the Universe is greater than the critical density, the geometry of space is closed and curved like the surface of a sphere. If the density of the Universe is less than the critical density, then the geometry of space is open (infinite), and curved like the surface of a saddle. If the density of the Universe exactly equals the critical density, then the geometry of the Universe is flat like a sheet of paper, and infinite in extent.

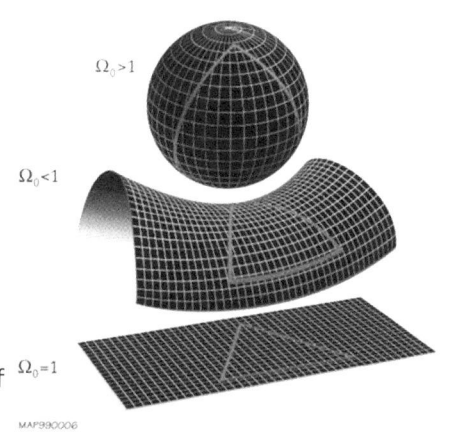

$\Omega_0 > 1$

$\Omega_0 < 1$

$\Omega_0 = 1$

MAP990006

Current thinking predicts that the density of the Universe is very close to the critical density, and that therefore the geometry of the Universe is flat. In case you're wondering what the critical density actually is, then it corresponds to roughly six hydrogen atoms per cubic metre – not really an awful lot to hang the fate of a Universe on!

Faster than light

Believe it or not, some galaxies are moving away from each other faster than the speed of light. How is this possible? Was Einstein wrong after all?

The important thing is that the galaxies aren't all rushing pell-mell through space. It's the Universe that's expanding and space itself that's getting bigger, and its carrying the galaxies along with it. So, although it's impossible to move through space faster than the speed of light, the rule doesn't apply to space itself, and it is indeed possible for the distances between galaxies to increase faster than the speed of light.

Could it, should it?

According to general relativity, the beginning of the Universe could have been in a Big Bang. But the question physicist Roger Penrose posed was: does general relativity predict that there should have been a Big Bang? Because saying something could have happened is not the same as saying that it did. In 1965, Penrose put together the way general relativity explains the behaviour of light cones with the fact that gravity is always an attractive force to demonstrate mathematically that a star collapsing under its own gravity would eventually become trapped in a region of space that shrinks to zero and contained within that zero volume the density of matter and the curvature of spacetime are infinite. It forms a black hole singularity.

At the same time that Penrose was working out his theorem, Stephen Hawking was looking for a subject for his PhD thesis. He read Penrose's work and realized that by reversing the direction of time in the theorem (scientifically a perfectly valid thing to do), and so instead of collapsing to zero, there is an expansion out from zero, the theorem still held. Penrose had shown that a collapsing star must end in a singularity. Hawking had demonstrated that if the current model of an expanding Universe was correct then it had to have begun with a singularity.

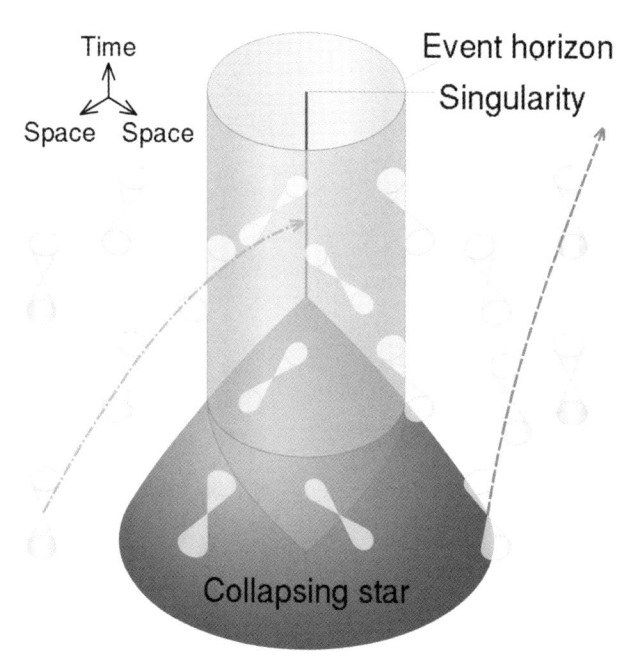

In 1970, Penrose and Hawking produced a joint paper which offered mathematical proof that if the description of the Universe given by Einstein's general theory of relativity was correct, and the Universe contains as much material as we observe that it does, then the Universe must have begun with a singularity.

Einstein's last words on the Big Bang, written years before Hawking and Penrose's work, were: 'One may... not assume the validity of the equations for very high density of field and matter, and one may not conclude that the "beginning of expansion" must mean a singularity in the mathematical sense.'

So far, Einstein's general theory of relativity has stood the test of time and experiment. There has been nothing to make scientists doubt its validity as a means of explaining the Universe as it is now. But we do know that it is an incomplete picture. It can't describe what happened at the start of the Universe because his theory predicts the breakdown of all physical laws, including itself, in the singularity. There must have been a time in the very early history of the Universe when events were dominated by the rules of that other great pillar of modern science – quantum mechanics.

Chapter 20

Does God play dice with the Universe?

In addition to relativity, another theory sent shockwaves through the world of physics as it entered the strange world of quantum mechanics.

For the last century, physics has been dominated by two great theories about how the Universe works. The practically simultaneous births of Max Planck's quantum theory in 1900 and Albert Einstein's relativity theory in 1905 marked the beginning of a period in which the very foundations of physical theory would be rebuilt.

Einstein's general relativity theory works on a grand scale, describing how gravity shapes the Universe of space and time. Quantum mechanics describes how the Universe works on the very smallest scales, down to the size of atoms and even smaller. The quantum realm is often described as an Alice in Wonderland world where events are mysterious, uncertain and inexplicable.

Both theories work extremely well. They have been tested by observation and experiment to extraordinary levels of accuracy and each one of them seems to reflect the Universe as it really is. The problem facing physics is that the two theories just don't join up. The laws of relativity

$$A(v) = \eta(v)\begin{pmatrix} 1 & v \\ \alpha v & 1 \end{pmatrix}$$

$$\eta(v) = \frac{1}{\sqrt{1-\alpha v^2}}$$

$$A'^0 = \frac{A^0 - \frac{v}{c}A^1}{\sqrt{1-\frac{v^2}{c^2}}}$$

$$E^2 - p^2c = m^2c^4$$

$$P = \frac{E}{c^2}u$$

$$\Delta t = \frac{u}{c^2}\Delta x$$

$$g_1 = \frac{(r\bar{b} + b\bar{r})}{\sqrt{2}}$$

$$g_4 = \frac{(r\bar{g} + g\bar{r})}{\sqrt{2}}$$

$$g_{\alpha\beta} = \begin{pmatrix} 1 & 0 & 0 & 0 \\ 0 & -1 & 0 & 0 \\ 0 & 0 & -1 & 0 \\ 0 & 0 & 0 & -1 \end{pmatrix}$$

$$\frac{v}{\sqrt{1-v^2/c^2}}$$

$$\Delta S^2 = c^2\Delta t^2 - \Delta x^2 - \Delta y^2 - \Delta z^2$$

$$v = v_0\sqrt{\frac{c-v}{c+v}}$$

$$c = \sqrt{c^2 + v^2}$$

$$F = \frac{ma}{\sqrt{1-u^2/c^2}} + \frac{mu\cdot(ua)/c^2}{(1-u^2/c^2)^{3/2}}$$

$$L = 2\pi R$$

$$g_6 = \frac{(b\bar{g} + g\bar{b})}{\sqrt{2}}$$

$$A_\alpha = g_{\alpha\beta}A^\beta = g_{\alpha 0}A^0 + g_{\alpha 1}A^1 + g_{\alpha 2}A^2 + g_{\alpha 3}A^3$$

that govern the Universe on the large scale don't apply on the small scale of quantum mechanics. The opposite is also true – quantum mechanics tells us nothing about the movements of galaxies or the geometry of the Universe. At the moment, there is no theory that will successfully combine gravity with quantum mechanics.

PLANCK AND EINSTEIN

Max Planck and Albert Einstein had huge respect and affection for each other. On the occasion of Planck's 60th birthday, Einstein spoke of Planck's 'inexhaustible persistence and patience' as he devoted himself to 'the most general problems of our science without letting himself be deflected by goals which are more profitable and easier to achieve. I have often heard that colleagues would like to attribute this attitude to exceptional will-power and discipline; I believe entirely wrongly so. The emotional state which enables such achievements is similar to that of the religious person or the person in love; the daily pursuit does not originate from a design or programme but from a direct need.' The same could probably have been said about Einstein himself.

Bohr's model

In 1913, Niels Bohr came up with a theory for the structure of an atom that was based on Planck and Einstein's ideas about quanta. He wanted to explain how it was that atoms could emit light quanta and also why it was that electrons didn't spiral down into the nucleus when they lost energy. To this end, he theorized that the electrons in an atom stay fixed distances from the atomic nucleus, arranged in orbits, or shells, around the atom. With his model, Bohr explained how electrons could jump from one orbit to another by emitting or absorbing energy in fixed quanta. For example, if an electron jumps one orbit closer to the nucleus, it must emit energy equal to the difference of the energies of the two orbits. Conversely, in order to jump to a higher orbit, the electron must

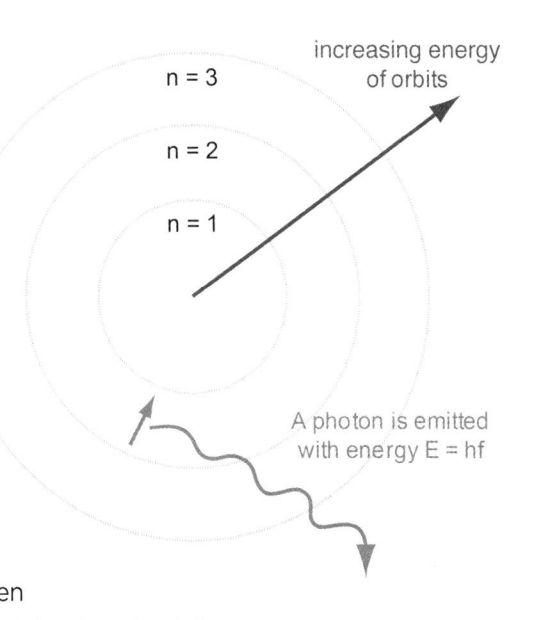

Lyman series

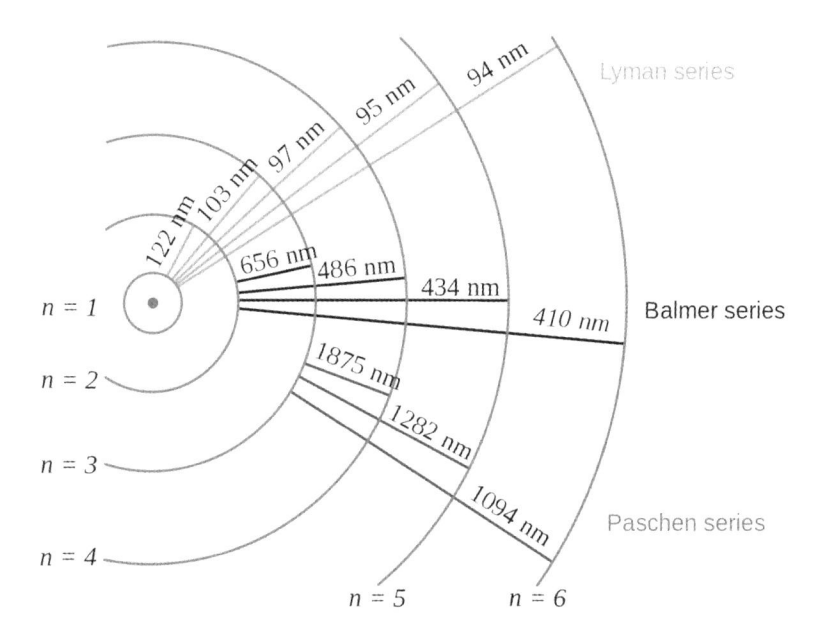

Balmer series

Paschen series

n = 1, n = 2, n = 3, n = 4, n = 5, n = 6

122 nm, 103 nm, 97 nm, 95 nm, 94 nm
656 nm, 486 nm, 434 nm, 410 nm
1875 nm, 1282 nm, 1094 nm

absorb a quantum of light equal in energy to the difference in orbits.

There were drawbacks to Bohr's theory, however. It worked well in describing the single electron of the hydrogen atom, but it ran into difficulties with larger atoms with multiple electrons and its assigning of a limited set of allowed orbits seemed somewhat arbitrary. It looked like a bit of a dead end. The new theory of quantum mechanics would resolve the difficulty.

Wave–particle duality again

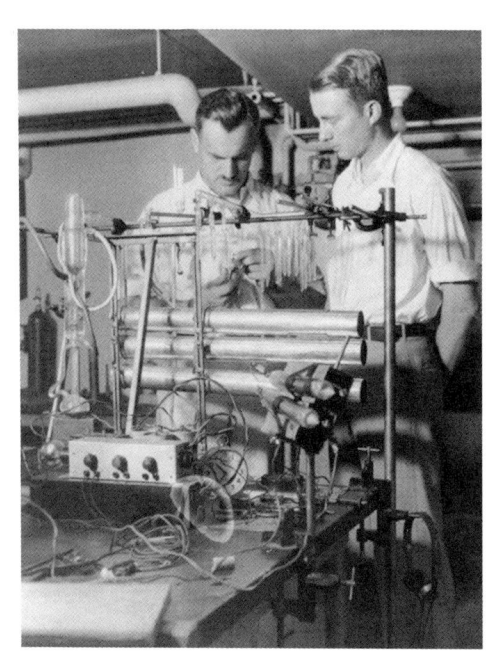

In his description of the photoelectric effect, Einstein had shown that light has both wave-like and particle-like properties. In an experiment carried out in 1922, American physicist Arthur Holly Compton highlighted the dual-wave-and-particle nature of electromagnetic radiation. The experiment involved sending a beam of X-rays through a target material. Compton observed that a small part of the beam was deflected off to the sides at various angles and that the scattered X-rays had longer wavelengths than the original beam. The change could be explained only by assuming that the X-rays were particles with discrete amounts of energy and momentum and applying conservation of energy and conservation of momentum laws to the collision between the photon and the electron. When X-rays are scattered, their momentum is partially transferred to the electrons they interact with. The electron takes some energy from an X-ray quantum, and as a result the frequency of the X-ray is reduced. The change in momentum and the frequency shift caused by the scattering are both explained by Einstein's quantum formula.

> ### WAVE–PARTICLE, PARTICLE–WAVE?
>
> In his 1932 doctoral thesis, French physicist Louis-Victor de Broglie proposed that not just light but all matter and radiation have both particle- and wave-like characteristics. Appealing to an intuitive belief in the symmetry of nature and Einstein's quantum theory of light, de Broglie asked, if a wave can behave like a particle then why can't a particle, such as the electron, also behave like a wave? De Broglie reasoned that as Einstein's famous $E = mc^2$ relates mass to energy and Einstein and Planck had related energy to the frequency of waves then combining the two suggested that mass should have a wave-like form as well. Einstein supported de Broglie's idea as it seemed a natural continuation of his own theories. Asked by the PhD board to comment on de Broglie's thesis he said:
> *'I believe de Broglie's hypothesis is the first feeble ray of light on this worst of our physical enigmas. It may look crazy but it is really sound!'*
> De Broglie got his doctorate.

Compton's experiment had settled the existence of photons, which until then had been disputed. Quantum mechanics followed within months.

Is there an electron there? Probably!

In the 1920s, researchers were studying the way a beam of electrons bounces off a piece of nickel. In this experiment, the nickel crystals acted in a similar way to the two slits used in the light interference experiment. Effectively this was the same experiment, but using a stream of electrons rather than a beam of light. The experiments found that the electrons formed an interference pattern just as the light did. The electrons were acting as if they were waves, as de Broglie had predicted. But what were these waves?

German physicist Max Born said that the wave was like a graph mapping out the probability of finding the electron in a particular place. Where the magnitude of the probability wave is large is where the electron is most likely to be found; if the magnitude is small then it is less likely the electron will be there. This sounds bizarre, and indeed it is one of the stranger notions in quantum mechanics. How can a particle be maybe here, perhaps there?

According to Born, the wave nature of matter means that everything has to be looked at in terms of probabilities. There is nothing hard and fast in the quantum realm. The best that we can

Wave-Particle Duality

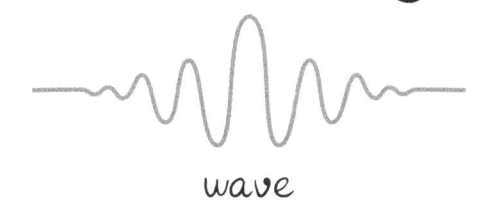

$c = 299\,792\,458 \text{ m/s}$

wave

particle

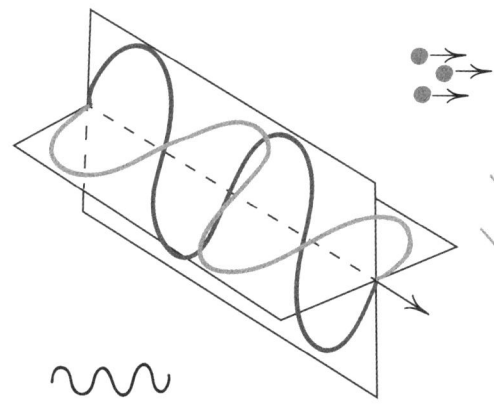

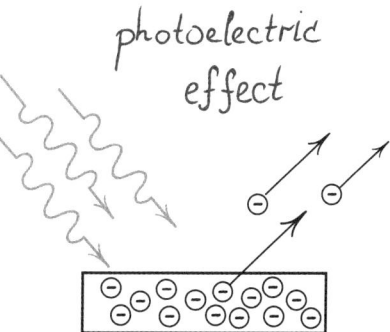

photoelectric effect

Does God play dice with the Universe? | **245**

ever do is to say the electron is likely to be somewhere, we can never say with certainty that it is there. The result of this is that you can carry out an experiment involving electrons and not get the same answer every time, even though you do everything the same way. All you can measure are probable outcomes, not certain ones.

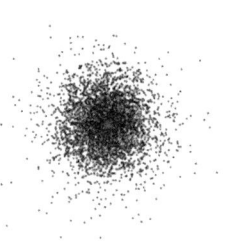

Heisenberg's uncertainty principle

Werner Heisenberg's famous uncertainty principle, first formulated in 1927, showed that it is impossible to know both the position and the momentum of a particle accurately at the same time. The more accurately the particle's momentum is measured the less accurately its position can be determined.

If it were possible to measure an electron's momentum with absolute precision its location would become completely uncertain – you might know how fast it was moving but you would have no clue where it was.

If the classical physicists

> **ON THE RIGHT WAVELENGTH**
>
> Even large objects, according to the formula worked out by de Broglie, have a wave-like nature. The de Broglie wavelength of an average car travelling at 40 km/h is around 6×10^{-38} m, in case you were wondering. It' a little too small to measure.

had been surprised by wave–particle duality then the uncertainty principle must have left them reeling. Everyday experience gives us no clue that such a thing is possible. If you're driving your car, for example, you have a good idea of its position and velocity (or at least you should). It's as if you knew you were doing 70 mph but had no idea if you were heading for Guildford or Glasgow.

These uncertainties have nothing to do with a lack of skill in the observer or inadequate equipment. Heisenberg showed that the uncertainty in the momentum multiplied by the uncertainty in the position of the particle can never be smaller than Planck's constant – it is a fundamental property of the Universe that puts a limit on what we can know.

Schrödinger's waves

In 1926, Austrian physicist Erwin Schrödinger worked out an equation that determines how these probability waves are shaped and how they evolve. The Schrödinger equation describes the form of the probability waves (or 'wavefunctions') that govern the motion of

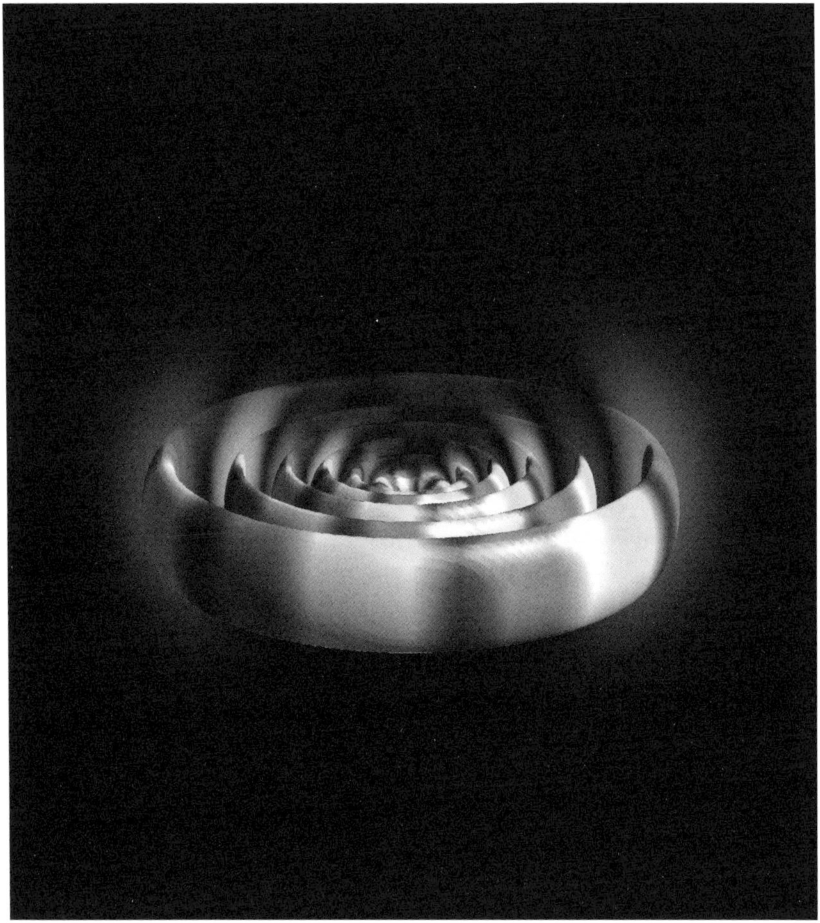

small particles, and it specifies how these waves are altered by external influences. Schrödinger tried out his equation on the hydrogen atom, and found that it predicted its properties with great accuracy.

It has been said that the Schrödinger equation is as important to quantum mechanics as Newton's laws of motion were for classical mechanics. Schrödinger was trying to describe the quantum world in mathematical terms, he wasn't trying to build up a model that you could picture in your head like the old idea of an atom as a mini solar system. Quantum mechanics was showing how the realm of the atom could be described in very precise and rigorous mathematical terms but with outcomes that could only be seen in terms of probabilities and not certainties.

According to quantum mechanics, when we perform a measurement to locate the position of a particle we cause its probability wave to collapse. It's no longer possible for it to be somewhere else when you know where it is – the probability of it being elsewhere drops to zero, while the probability of it being where you observed it rises to 100 per cent.

Even today, there is still disagreement as to whether or not a wavefunction is a real physical thing or just a mathematical tool that allows us to calculate quantum realm probabilities but with no basis in reality. From a practical perspective, it doesn't seem to matter. The Copenhagen interpretation of quantum theory, developed in the 1920s, principally by physicists Niels Bohr and Werner Heisenberg, treats the wavefunction as no more than a tool for predicting the results of observations, and says that physicists shouldn't concern themselves with trying to imagine what 'reality' looks like. It's an approach that physicist David Merman memorably summed up as 'Shut up and calculate!' The certainties of the clockwork Universe of Newton's day had been left a long way behind.

The Copenhagen interpretation

In the Copenhagen interpretation of quantum mechanics, championed by Niels Bohr and others (named after the city where Bohr was based), the properties of a quantum particle have no definite value until a measurement is made. The principle of complementarity is central to the Copenhagen interpretation. This says that the wave and particle nature of objects are complementary aspects of a single reality, like the two sides of a coin. An electron or a photon, for example, can behave sometimes as a wave and sometimes as a particle, but never both at the same time, just as a tossed coin can be either heads or tails, but not both simultaneously.

THE SOLVAY CONFERENCES

Instituted by Belgian industrialist Ernest Solvay, the Solvay Conferences on physics and chemistry are held in Brussels. The first physics conference took place in 1911 and the first chemistry conference in 1922. They keep to a three-year schedule, with a physics conference in the first year, no conference in the second year, and a chemistry conference in the third year. The fifth physics conference in 1927 is famous for the part it played in formulating the ideas that would come to dominate quantum mechanics and for the dispute between Albert Einstein on the one hand and Niels Bohr, Werner Heisenberg and Max Born on the other regarding the validity of the Copenhagen interpretation.

Niels Bohr said it was meaningless to ask what an electron really is. Experiments designed to measure waves will see waves, while experiments designed to measure particle properties will see particles. It is impossible to design an experiment that would allow us to see wave and particle at the same time. The wavefunction is a complete description of a wave/particle. When a measurement of the wave/particle is made, its wavefunction collapses. Any information that cannot be obtained from the wavefunction does not exist.

Particle waves, according to Max Born in the 1920s, are measures of probability. They're not physical entities like sound or water waves. We can never be absolutely certain as to how any given particle will behave; identical electrons might do different things every time an experiment is run, so the outcome of an experiment can only be predicted statistically.

The Copenhagen interpretation of quantum

mechanics opened a sharp divide between classical Newtonian physics and quantum physics. Here, in the world of everyday things, we expect every event to have a cause. The glass of water on the table didn't fall over by itself, it was knocked over when you stumbled and bumped into the table. We might not have predicted you were going to do that but if we'd had all the details, such as your stride length, and the height of the wrinkle in the rug you caught your foot on, we could have come up with a reasonable probability of the spillage occurring. So, according to classical physics, all the variables are there, even if sometimes they might be hard to measure.

In the quantum world, however, there are no relevant details to take into account, there is just pure chance. The quantum world is pure statistical probability. The Copenhagen view is that indeterminacy is a fundamental feature of nature and not just a result of our lack of knowledge. We just have to accept that this is how things are and not try to explain it.

Einstein speaks out

Some physicists worried about this lack of explanation, among them Albert Einstein. In the spring of 1927, the 200th anniversary of Newton's death, two decades after Einstein had, with seeming ease, overturned much of classical physics with special relativity, he came to the defence of classical mechanics and causality. 'The

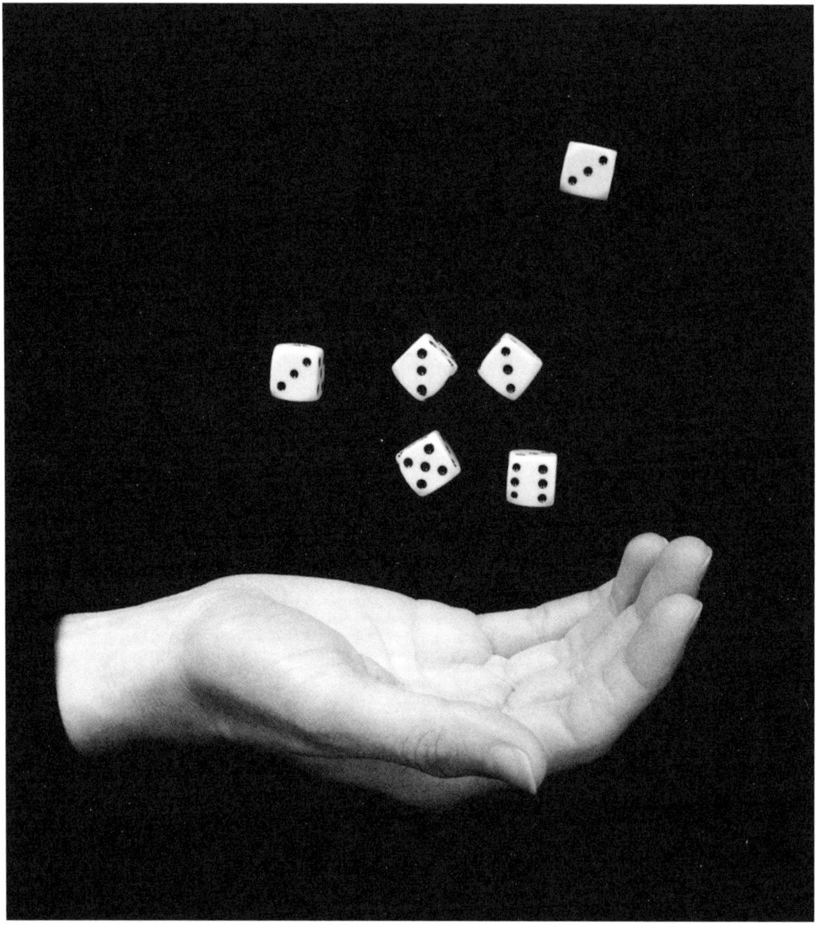

last word has not been said,' Einstein argued. 'May the spirit of Newton's method give us the power to restore union between physical reality and the profoundest characteristic on Newton's teaching – strict causality.'

Einstein never came to terms with quantum theory, believing that it was correct as far as it went, but fundamentally incomplete. He couldn't accept a reality that was defined by uncertainty, probability and indeterminacy. He believed that the probabilities of quantum mechanics resulted from a gap in our knowledge of how the Universe operated on the atomic scale. Once we had a fuller understanding, probability would be replaced by certainty.

Einstein once said to a friend that when he was judging a theory he asked himself whether, if he were God, 'I would have arranged the world in such a way'. He couldn't believe that there were rules that governed most of what happened in the Universe but at the fundamental quantum level of reality things seemed to be left to chance.

In a letter to Max Born, written in 1926, Einstein stated: 'Quantum mechanics is certainly imposing. But an inner voice tells me that it is not yet the real thing. The theory says a lot, but does not really bring us any closer to the secret of the "old one". I, at any rate, am convinced that He does not throw dice.'

EINSTEIN AND THE LASER

Among all his other achievements, it is perhaps not so widely known that Einstein also contributed to the development of the laser. 'Laser' is an acronym for 'Light Amplification by Stimulated Emission of Radiation'. It is a device that creates and amplifies a narrow, focused beam of light, and it can trace its roots back to a 1917 paper by Albert Einstein on the quantum theory of radiation.

In a laser, atoms or molecules of either a crystal such as ruby or garnet, or a gas or liquid are 'pumped' to push them up to higher energy levels. This produces a burst of light as the atoms discharge a flood of photons. This is called stimulated emission, a process Einstein first suggested as a possibility in his 1917 paper. After completing work on the general relativity theory the previous year, he had turned to exploring the interplay of matter and radiation. It was in the course of this work that he came up with an improved fundamental statistical theory of heat that included the quantum of energy.

Einstein proposed that an excited atom, one that has absorbed a photon, can return to a lower energy state by re-emitting the photon, a process he called spontaneous emission. He also predicted that as light passes through a substance, it could stimulate the emission of more light. His idea was that if you had a large number of atoms in an excited state, all ready to emit a photon, a stray photon passing by could stimulate them to release their photons. These released photons would have the same frequency and direction as the original photon. A cascade of photons are released as the identical photons move through the rest of the atoms. Of course, Einstein never put his theory into practice – it would have to wait until 1960 when the first working laser was built.

Chapter 21

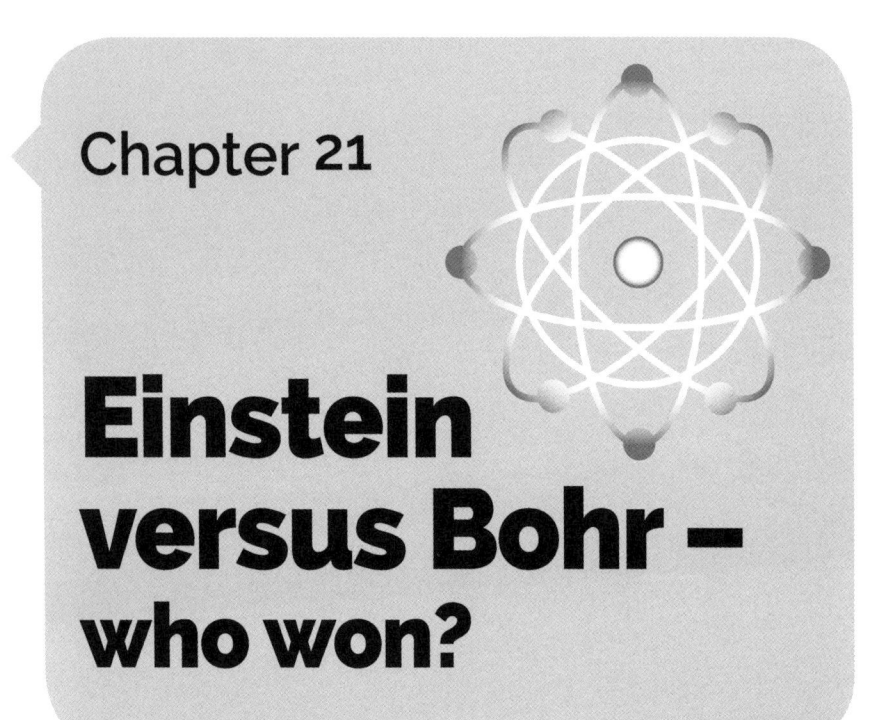

Einstein versus Bohr – who won?

Albert Einstein and Niels Bohr debated the merits and shortcomings of quantum theory back and forth for decades. Could either man be said to have won the argument?

Over a period of three decades, until Einstein's death, Einstein and Bohr repeatedly challenged each other's beliefs and interpretations of the quantum world. These debates were never acrimonious – the two physicists were great friends – but each man held to his point of view and defended it stubbornly. Einstein believed that there was an objective reality that existed and could be measured, whereas Bohr believed that the very act of measurement altered reality. For example, an electron has no definite position in space until someone decides to measure it.

The Einstein–Bohr contest

Einstein and Bohr crossed paths a number of times during the course of the fifth Solvay conference on physics. Einstein understood perfectly well what the quantum mechanics were proposing but felt that what they were presenting just wasn't the complete picture. Bohr, who had been confident that Einstein would come on board with the Copenhagen interpretation, was shocked and dismayed by Einstein's opposition.

After the conference, Einstein and Bohr entered into a series of contests in which Einstein would attempt to find flaws in Bohr's interpretation of quantum mechanics and Bohr would defend his stance. Einstein would present Bohr with a thought experiment and Bohr would find a flaw in Einstein's argument, usually within a few days.

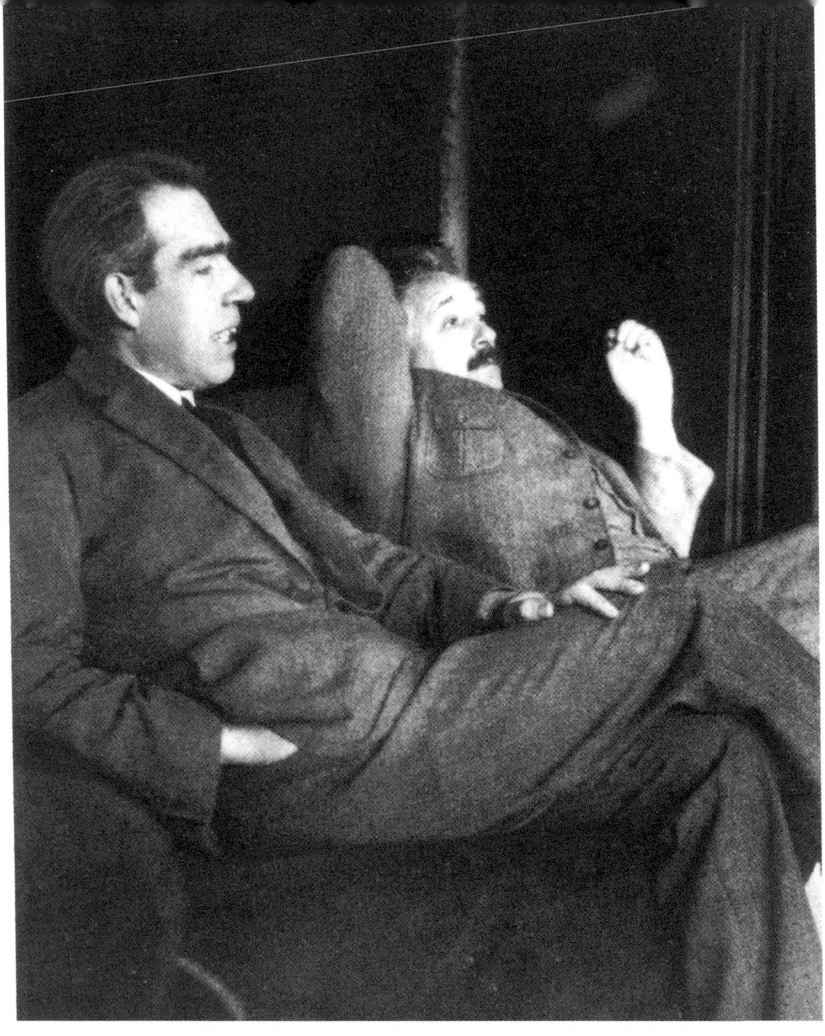

In 1948, Bohr summarized the discussions he had had with Einstein up until then. He concluded: 'Whether our meetings have been of short or long duration, they have always left a deep and lasting impression on my mind...'

A box full of light

One of the most celebrated contests between Einstein and Bohr was one in which Einstein asked Bohr to imagine a box full of light. The box had a number of clocks and scales set up inside it and these could be used to determine both the energy and the time of release of a single photon. First the box had to be weighed, then a single photon would be released through a shutter operated by a clockwork mechanism inside the box. The box could then be weighed again and, knowing the change in mass, Einstein could calculate the energy of the photon using $E = mc^2$. He would therefore know the change in energy as well as the precise time the photon was emitted, thus evading the uncertainty principle.

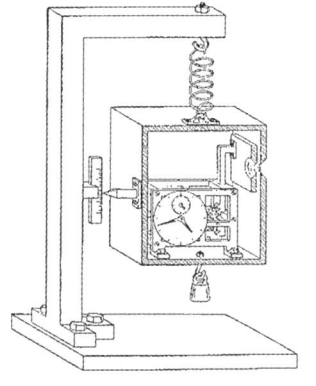

By all accounts, Bohr spent a sleepless night trying to come up with a riposte to Einstein's light box. Then the answer came to him. The photon, he realized, would have a recoil when it was fired into the box, thus causing uncertainty about the position of the clock in Earth's gravitational field, and since Einstein himself in the general theory of relativity had shown that clocks run slower in a gravitational field there would be uncertainty in the time recorded by the clock. Einstein had caught himself out by forgetting about his own theory!

Spooky action at a distance

In 1935, Einstein, in collaboration with colleagues Boris Podolsky and Nathan Rosen, introduced another thought experiment that argued that quantum mechanics was not a complete physical theory. Known today as the 'EPR paradox' after the three collaborators, the thought experiment was meant to tackle a feature of quantum mechanics called quantum entanglement. This said that the result of a measurement on one particle of an entangled quantum system can have an instantaneous effect on another particle, regardless of the distance between the two parts.

As we have seen, one of the main tenets of quantum mechanics is the idea of uncertainty – we can't measure all of the features of a system simultaneously, not even in theory. We can't know, for example, position and momentum and have to choose to measure one or the other, but not both together. Another peculiar property of quantum mechanics is called entanglement. Two photons, for example, are allowed to interact so that they can subsequently be defined by a single wavefunction. (How this is accomplished need not detain us.) Once they are separated, they will still share this single wavefunction, which means that measuring one will determine the state of the other: for example, if the two quanta have a spin-zero entangled state then, if one particle is measured to be in a spin-up state, the other is instantly forced to be in a spin-down state. This is officially known as 'non local behaviour'. Einstein called it 'spooky action at a distance'.

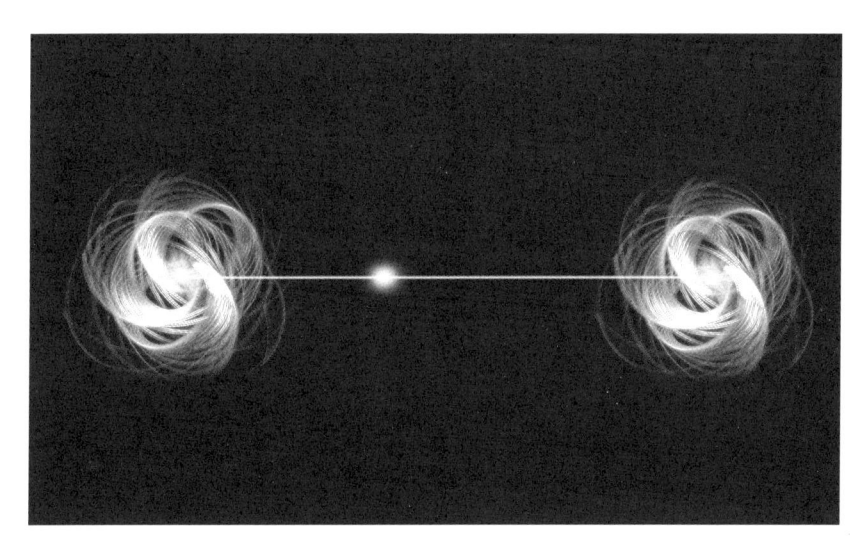

Einstein accepted that quantum mechanics could accurately

> ## SPIN
>
> In the 1920s, Otto Stern and Walther Gerlach carried out a series of important experiments at the University of Hamburg. Knowing, as we do, that all moving charges produce magnetic fields, they wanted to measure the magnetic fields produced by the electrons orbiting nuclei in atoms. The physicists were surprised to find that the electrons themselves act as if they are spinning very rapidly, producing tiny magnetic fields independent of those resulting from their orbital motions. The term 'spin' was soon used to describe this apparent rotation of subatomic particles. This shouldn't be taken to mean that electrons are actually small, solid bodies spinning in atomic space – they're not.

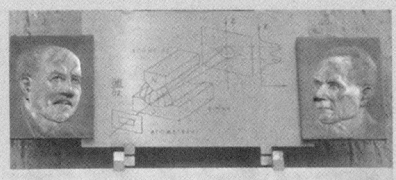

predict the outcomes of various experiments; he knew that it wasn't 'wrong'. What he was doing was arguing that it wasn't yet complete and the EPR paradox was another attempt to demonstrate this – the paper was even entitled, 'Can the Quantum Mechanical Description of Physical Reality Be regarded as Complete?' Einstein suggested that there were properties of the quantum system that remained to be discovered, what he called 'hidden variables', which once known would account for the observations and mean that there need be no recourse to 'spooky action'. Bohr, naturally, disagreed with Einstein's view and passionately defended the Copenhagen interpretation of quantum mechanics.

Einstein and his co-authors began by setting out their premise, that if there was any way that we could learn with absolute certainty the position of a particle, and we don't disturb the particle by directly observing it, then we can say the particle exists in reality, independent of our observations.

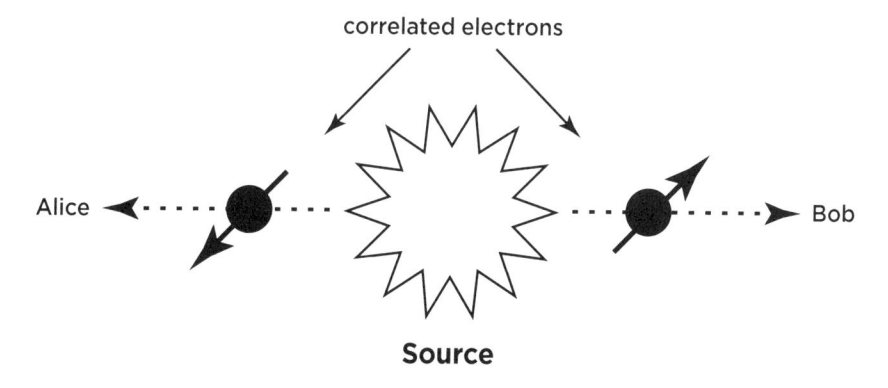

If we have two quantum entangled particles then we can take measurements of one particle that give us information about the second particle without disturbing the second particle in any way. By measuring the momentum, say, of the first particle, we gain precise knowledge of the momentum of the second particle and we can do the same for other properties, such as position.

So, the second particle, which we have not directly observed, has properties that we know. It has a position that is real and a momentum that is real. Since quantum mechanics tells us that we can't know both these properties then it appears that quantum mechanics' description of reality is indeed incomplete.

The alternative, Einstein and Co argued, was to assume that the process of measuring the first particle alters the reality of the second, instantaneously making it conform to the reality of the first particle, even if they were separated by light years of space. 'No reasonable definition of reality could be expected to permit this,' they asserted.

Wolfgang Pauli made his feelings very plain in a letter to Werner Heisenberg:

'Einstein has once again expressed himself publicly on quantum mechanics (together with Podolsky and Rosen – no good company by the way). As is well known, every time that happens it is a catastrophe.'

When the EPR paper reached Niels Bohr, he knew that he would have to find a riposte to

Einstein. According to one of Bohr's colleagues in Copenhagen, EPR came on them like 'a bolt out of the blue... Everything else was abandoned. We had to clear up such a misunderstanding at once.' It was no easy task. It took six weeks of fretting before Bohr was finally ready to make his response. It was longer than the four-page EPR paper had been.

Bohr admitted that in Einstein's paper 'there is no question of a mechanical disturbance of the system under investigation'. Until then, Bohr had asserted that the disturbance caused by making a measurement of a particle was what led to quantum uncertainty. He backed away from that position now. In various arguments at the Solvay conferences he had often rejected Einstein's thought experiments by recourse to the uncertainty principle. Now, however, he used the concept of complementarity. The most important aspects of a quantum experiment were the conditions under which it was made. If you chose one set of conditions, for example an experiment involving wave properties, then wave properties were what you would see. If you chose something else then you would reveal a complementary aspect to the wave properties. None of these elements, Bohr felt, were present in the EPR thought experiment and so it failed to refute the Copenhagen interpretation of quantum mechanics.

If the two particles are entangled, Bohr argued, then they are effectively a single system that has a single quantum function. In addition, he noted, the EPR paper did not absolutely overrule the uncertainty principle. It still is not possible to know both the precise position and the precise momentum of the particle at the same moment. If you know the position of A then you know the position of B, and if you know the momentum of A you know the momentum of B. But what you still can't do is know these two things precisely at the same moment for A, so you can't know them for B either. There is no conflict with the uncertainty principle.

Einstein continued to insist that he was on to something. His own theory of relativity did not permit 'spooky action at a distance'. He had forbidden it for Newton's gravity and he wasn't going to allow it for quantum mechanics. Quantum mechanics, he maintained, violated two fundamental principles. The principle of separability, which maintains that two systems separated in space have an independent existence; and the principle of locality, which says that doing something to one system cannot immediately affect the second system.

Einstein's box, Schrödinger's cat

Erwin Schrödinger was among those who sided with Einstein in opposing the Copenhagen interpretation. He said of the EPR paper: 'Like a pike in a goldfish pond it has stirred everyone up.' He felt his wave equations had been misused and sometimes thought it might have been better if he had never developed them. At one point he declared of quantum mechanics, 'I don't like it, and I'm sorry I ever had anything to do with it.' In a letter to Schrödinger in 1928, Einstein complained: 'The Heisenberg–Bohr tranquillizing philosophy … provides a gentle pillow for the true believer from which he cannot very easily be aroused.'

Einstein thought that Heisenberg's uncertainty principle might be a demonstration of the limits nature places on what we can know about a quantum object but, none the less, these limits should not be taken to imply that there wasn't a deeper, more deterministic reality, only that we were denied access to it.

In 1935, Einstein shared a thought experiment with Schrödinger that illustrated why he felt so uncomfortable with wavefunctions and probabilities. Imagine two boxes, he said, one contains a ball, the other is empty. Before we look in a box there is a 50 per cent chance of finding the ball. After we look, the chance of it being there is either 100 per cent or 0 per cent. But in reality, the ball was always 100 per cent in one of the boxes. Einstein wrote:

'… the probability is ½ that the ball is in the first box. Is that a complete description? NO: A complete description is that the ball is (or is not) in the first box… YES: Before I open them, the ball is by no means in one of the two boxes. Being in a definite box comes about only when I lift the covers.'

Clearly Einstein favoured the first answer and not the quantum mechanical second answer. Niels Bohr and the Copenhagen interpretation would say that the ball exists in a state of superposition, that it actually occupies both boxes until you look and see which one it's in. The act of observation makes the choice. Einstein's answer is based on common sense, but as he had demonstrated himself with his relativity theories, common sense is not always a reliable guide to how nature actually works.

   Schrödinger came up with a thought experiment of his own, which would pass into quantum folklore. It examined a core concept in quantum physics, which was that the timing of the emission of a neutron from a decaying nucleus is indeterminate until it is observed. In the quantum world the nucleus exists simultaneously in both its decayed and undecayed states, until observation collapses its wavefunction and it becomes either one or the other. It is a state of affairs that we might reluctantly accept as being true in the strange realm of the quantum but how on Earth can these odd goings-on be scaled up into the 'real' world?

   In his thought experiment, Schrödinger posed the following question: when does the system switch from its state of superposition into one definite reality? Enter the cat.

   'A cat is penned up in a box,' Schrödinger wrote, 'along with the following device: in a Geiger counter there is a tiny bit of radioactive

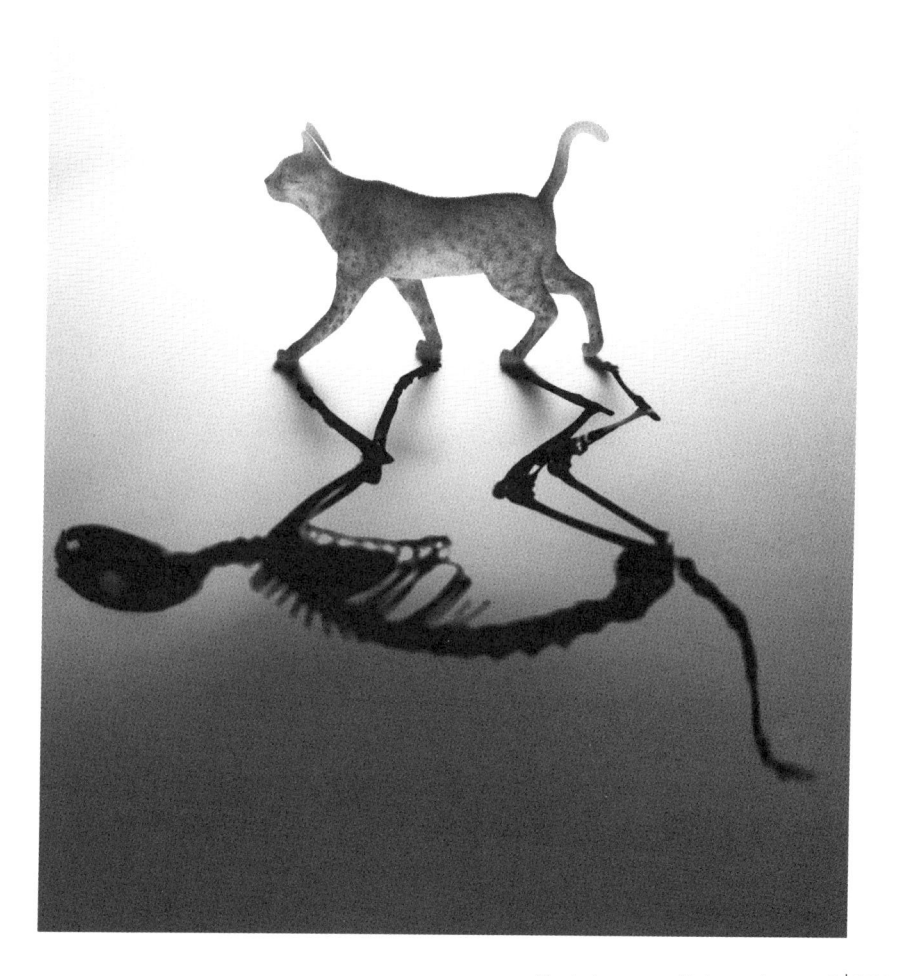

Einstein versus Bohr – who won? | **271**

substance, so small, that perhaps in the course of the hour one of the atoms decays, but also, with equal probability, perhaps none: if it happens… a relay releases a hammer which shatters a small flask of hydrocyanic acid.'

He explained that the wavefunction of the entire system would express the situation by having in it the living or dead cat 'mixed or smeared out'. Einstein and Schrödinger were happy that their thought experiments had demonstrated their point – there was something distinctly not right about the Copenhagen interpretation. Einstein said that a wavefunction that 'contains the living as well as the dead cat just cannot be taken as a description of a real state of affairs.'

In 1948, Einstein wrote to Max Born, saying: 'You believe in a dice-playing God and I in perfect laws in the world of things existing as real objects, which I try to grasp in a wildly speculative way.' For Niels Bohr, on the other hand, there was no reason why the rules of classical physics, which determine what goes on in the everyday world around us, should also apply to the quantum realm. What the quantum physicists were discovering was just the way things were, whether Einstein liked it or not. At some point, Bohr apparently said exasperatedly to Einstein: 'Stop telling God what to do!'

Born, expressing a disappointment felt by many physicists, said of Einstein that he was 'a pioneer in the struggle of conquering the wilderness of quantum phenomena. Yet later, when out of his own work a synthesis of statistical and quantum principles emerged which seemed acceptable to almost all physicists he kept himself aloof and skeptical. Many of us regard this as a tragedy – for him, as he gropes his way in loneliness, and for us, who miss our leader and standard bearer.'

Einstein never accepted the probabilities and uncertainties of quantum mechanics and sought throughout his life to find an underlying order. None the less, quantum mechanics has stood up to experiment in the years since Einstein's death, and all the indicators are that he was wrong. As Stephen Hawking commented in a 1997 lecture: 'Einstein was confused, not the quantum theory.'

Chapter 22

Was Einstein the 'father of the atomic bomb'?

Popular imagination links Einstein to the creation of the atomic bomb, but how big a role did he really play in its development?

The popular imagination almost inevitably links Albert Einstein and $E = mc^2$ with the invention of the atomic bomb, but how big a role did he really play in its development? The first use of the term 'father of the bomb' was probably in an article in *Time* magazine, which featured an image of Einstein against a mushroom cloud reading '$E = mc^2$', on its 1 July 1946 cover.

Discovering the atom

Around the time that Albert Einstein was working on general relativity, Ernest Rutherford was exploring the structure of the atom at the Cavendish Laboratory in Cambridge, England. In 1907, he devised an experiment to show that the atom had a centre, which he called the nucleus. Remember, this was just two years after Einstein's paper on Brownian motion had confirmed the existence of atoms. By 1919, the year Arthur Eddington's eclipse observations had confirmed general relativity, Rutherford had succeeded in transforming atomic nitrogen into hydrogen, or, as the papers put it, 'split the atom'.

One of Rutherford's students was Niels Bohr, who would play a large role in Einstein's life. It was Bohr who formulated the model of atomic structure that explained the release of photons of different energies, which dovetailed with Einstein's idea that light was a stream of particles.

Scientists began to find evidence that Einstein's $E = mc^2$ equation

was indeed right. Francis Ashton, a fellow researcher at the Cavendish, made careful measurements of the atomic weights of the elements and was surprised to discover that there was a tiny amount of mass missing. This he believed was the energy that held the atoms together, which he called binding energy. He calculated that if it were possible to transform hydrogen, the lightest element, into helium, the next lightest, 1 per cent of the mass would be annihilated and released as energy. According to Einstein's formula, there would be enough energy in a glass of water to power a steamship across the Atlantic and back again.

It was the first time Einstein's equation had been linked with atomic research. But how could scientists gain access to that vast, untapped store of energy? Many thought it unlikely, including Rutherford, who in a speech in 1933 dismissed the idea as 'merest moonshine'. As it turned out, ironically, the attempts to unlock the energy of the atom began to show promise just as the world stood on the brink of war.

The journey to the bomb

A crucial breakthrough had been made in Cavendish's laboratory a year before he made his 'moonshine' comment. This was the discovery of the neutron in the atomic nucleus. Neutrons are highly penetrative particles and if atomic nuclei were to be shattered to release their energy neutrons would be the sparks that lit the fire.

By 1934, Irene Curie, daughter of Marie, had successfully created a new radioactive element; in that same year, Enrico Fermi in Rome demonstrated that neutrons could be made more effective atom smashers by slowing them down. In 1938, Otto Hahn, working in Berlin, was baffled when he bombarded uranium with neutrons and found he was left with barium. In collaboration with Austrian physicist Lise Meitner and her nephew, Otto Frisch, Hahn realized that in splitting the uranium atom he had released some of its binding energy. This was the first successful demonstration of nuclear fission.

News soon spread through the physics community. One scientist who heard the news, Leo Szilard, a Hungarian living and working in New York, was a friend of Einstein. Szilard had obtained his doctorate in physics in Berlin with Einstein's help. He was a theoretician and since 1933 had been researching the possibility that an atom split by a neutron might release two more neutrons, thereby triggering a chain reaction. Together with Fermi, he showed this to be the case in March 1939.

In Princeton, Niels Bohr had determined that it was the U-235 isotope of uranium that was easiest to fission when bombarded with neutrons. The problem was that this made up less than 1 per cent of natural uranium and it would be very difficult to separate it out. If it could be done though, Bohr warned, it would be possible to set off a devastating atomic explosion.

In Germany, meanwhile, physicists were pursuing the same ideas

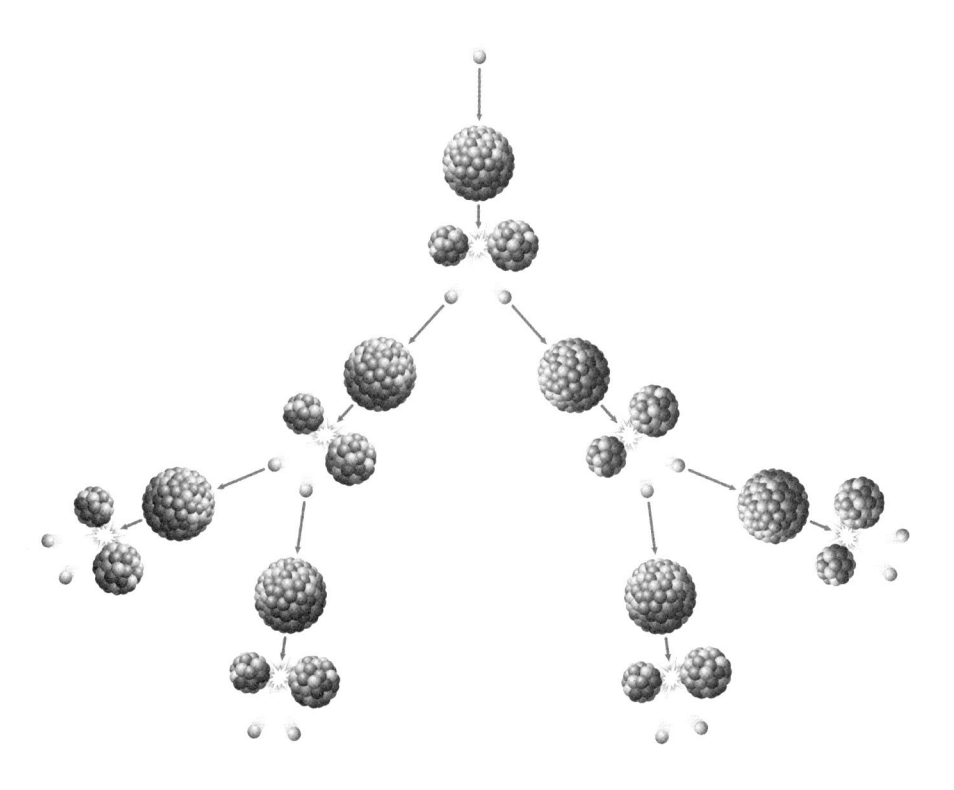

EINSTEIN'S REFRIGERATOR

In 1930, Einstein and Leo Szilard set themselves the task of devising a noiseless household refrigerator. Part of their invention was the so-called Einstein–Szilard pump, later described by Einstein as using an alternating electric current to generate a magnetic guide field that moves a liquid mixture of sodium and potassium. The mixture, according to Einstein, 'moves in alternating directions inside a casing and acts as the piston of a pump; the refrigerant [inside the casing] is thus mechanically liquified and cold is generated by its re-evaporation.' The inventive physicists seemingly received a small amount of money for their work, certainly not enough to make them rich. The Einstein–Szilard refrigerator was never marketed commercially, partly because of the dangers from leakage of the poisonous refrigerant.

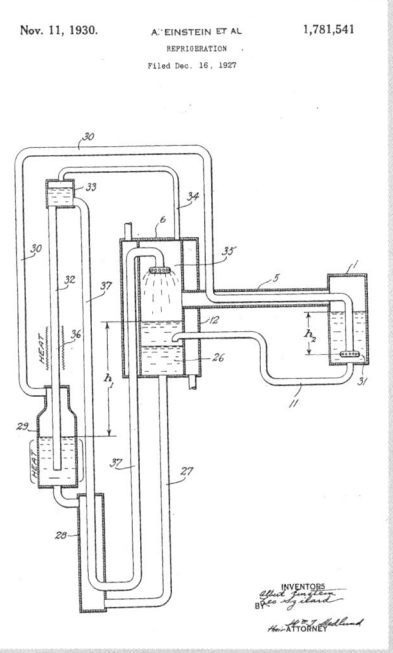

as Szilard. Worried by the implications of this, Szilard pushed for a security embargo on the reporting of all nuclear research in the United States, Great Britain, France and Denmark. He was right to be worried, as the Nazis embarked on a fission research programme in April 1939.

When word of the discovery of nuclear fission reached Robert Oppenheimer, who would go on to mastermind the Manhattan Project, the code name for the top secret development of the atom bomb by the Allied powers in the Second World War, he at first declared it to be 'impossible'. After it was demonstrated to him that the experiment did indeed work, he too immediately began investigating chain reactions. Within days, he had drawn up a crude plan for an atomic bomb.

Einstein and Roosevelt

On 2 August 1939, Leo Szilard visited Einstein and urged him to write to President Roosevelt and impress on him the need to begin work on the development of atomic weapons. This visit led to the letter signed by Einstein, though probably largely written by Szilard, that was sent to Roosevelt on 11 October 1939. In it Einstein warned of the possibility of achieving a chain reaction in uranium in the near future, generating vast amounts of power.

'This new phenomenon,' Einstein wrote, 'would also lead to the

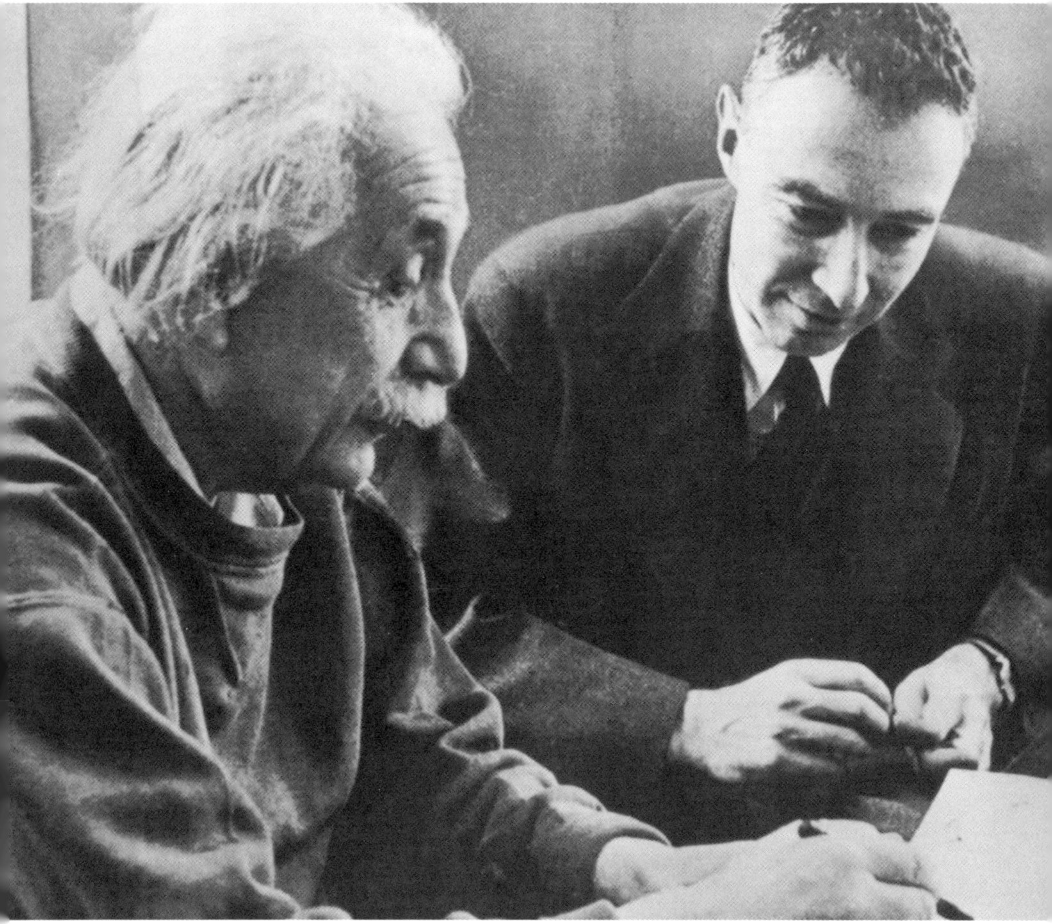

construction of bombs, and it is conceivable – though much less certain – that extremely powerful bombs of this type may thus be constructed. A single bomb of this type, carried by boat and exploded in a port, might very well destroy the whole port together with some of the surrounding territory. However, such bombs might very well prove too heavy for transportation by air.'

He went on to suggest that Nazi Germany might already be carrying out such research.

Roosevelt replied that he had found the data 'of such import that I have convened a Board… to thoroughly investigate the possibilities of your suggestion regarding the element of uranium.'

According to Einstein's biographer Abraham Pais, opinions are divided as to how much influence Einstein's letter to Roosevelt actually had. It was Pais's impression that it was marginal, pointing out that, although Roosevelt did appoint an advisory committee, it wasn't until October 1941 that he gave the go-ahead for full-scale atomic weapons development and 6 December 1941, the day before the Japanese attack on Pearl Harbor, that the Manhattan Project was launched.

Even if he had played a part in setting it in motion, Einstein never worked directly on the Manhattan Project to develop the atomic bomb. He was not invited to join it nor even officially told that it existed. J. Edgar Hoover, the director of the FBI, was suspicious of Einstein's pacifism and believed him to be a security risk. Among other things, Hoover asserted that Einstein had supported an anti-war congress in 1932 and was pro-Soviet, whereas, in fact, Einstein had refused to attend the conference and had denounced Russia for 'its complete suppression of the individual and of freedom of speech'.

Einstein did play a small role in the Manhattan Project, however. He was asked by Vannevar Bush, one of the project's chief scientists, to help with a problem involving the separation of isotopes. Einstein worked for two days on a process in which uranium was converted into a gas and forced through filters and sent in his report. The scientists who looked at it were keen for Einstein to be given a bigger role in the project but Bush

> **EINSTEIN'S AUCTION**
>
> To help the war effort, Einstein submitted a copy of the 1905 paper on special relativity for auction. It wasn't the original – Einstein had thrown that away after it was published! To recreate the manuscript, he had his assistant, Helen Dukas, read it to him aloud while he wrote down the words. The manuscript, along with one other, sold for $11.5 million.

refused. 'I wish very much that I could place the whole thing before him and take him fully into confidence,' Bush wrote, 'but this is utterly impossible in view of the attitude of people here in Washington.'

In December 1944, Einstein was visited by his friend Otto Stern, who had been working on the Manhattan Project. Stern's news that the project was nearing completion upset Einstein. He decided to write to Niels Bohr about his concern for the control of atomic weapons in the future. 'The politicians do not appreciate the possibilities and consequently do not know the extent of the menace,' he wrote. Bohr visited Einstein and urged caution in making his views more widely known, warning of 'the most deplorable consequences' if information about the development of the bomb were to get out. Einstein agreed.

On 6 August 1945, an atom bomb obliterated the Japanese city of Hiroshima. Einstein heard the news at a cottage he rented in the Adirondacks. All he said was 'Oh, my God.' A few days later, following the dropping of a second bomb on Nagasaki, a report was issued detailing the development of the bomb. Much to Einstein's dismay, the report laid great emphasis on his letter to Roosevelt. It was one of the reasons why popular imagination linked Einstein to the bomb, even though he had taken little part in its development.

After the war

In an interview with *Newsweek* magazine, Einstein declared, 'Had I known that the Germans would not succeed in producing an atomic bomb I never would have lifted a finger.' In December 1945, Einstein told an audience that 'The first atomic bomb destroyed more than the city of Hiroshima. It also exploded our inherited, outdated political ideas.' He was the chairman of the Emergency Committee of Atomic Scientists, a group that met between 1946 and 1949. In the group's charter, Einstein stated his belief in the need 'to advance the use of atomic energy in ways beneficial to mankind [and] to diffuse knowledge and information about atomic energy… in order that an informed citizenry may intelligently determine and shape its action to serve its own and mankind's best interest.'

Chapter 23
Can we find a theory of everything?

Many theorists today are looking for a way to unite relativity and quantum mechanics. Are they close to finding an answer?

Einstein's relativity theories provide a framework for understanding the Universe on the scale of stars and galaxies; quantum theory describes how the Universe works on the scale of atoms and atomic particles. Both theories have been tested to unimaginable levels of accuracy, both appear to work, but the problem remains that we have yet to find a way to unite the two. If quantum mechanics is used together with general relativity to calculate the probability of a process involving gravity taking place the answer that comes up is an infinite probability. This is not a good thing. Any answer above 100 per cent probability is effectively meaningless. It just demonstrates that combining general relativity and quantum mechanics simply doesn't work.

Einstein spent the last 30 years of his life trying to find a way to unite electromagnetism and gravity but never succeeded. He was convinced that there had to be a single theory that would encompass all of the Universe's physical phenomena. In his Nobel Prize acceptance lecture he said: 'The intellect seeking after an integrated theory cannot rest content with the assumption that there exist two distinct fields totally independent of each other by their nature.'

When Einstein began his work on a unified field theory in the 1920s, electromagnetism and gravity were the only known forces, and the only subatomic particles that had been discovered were the electron and the proton. Now physicists have learned that there are two

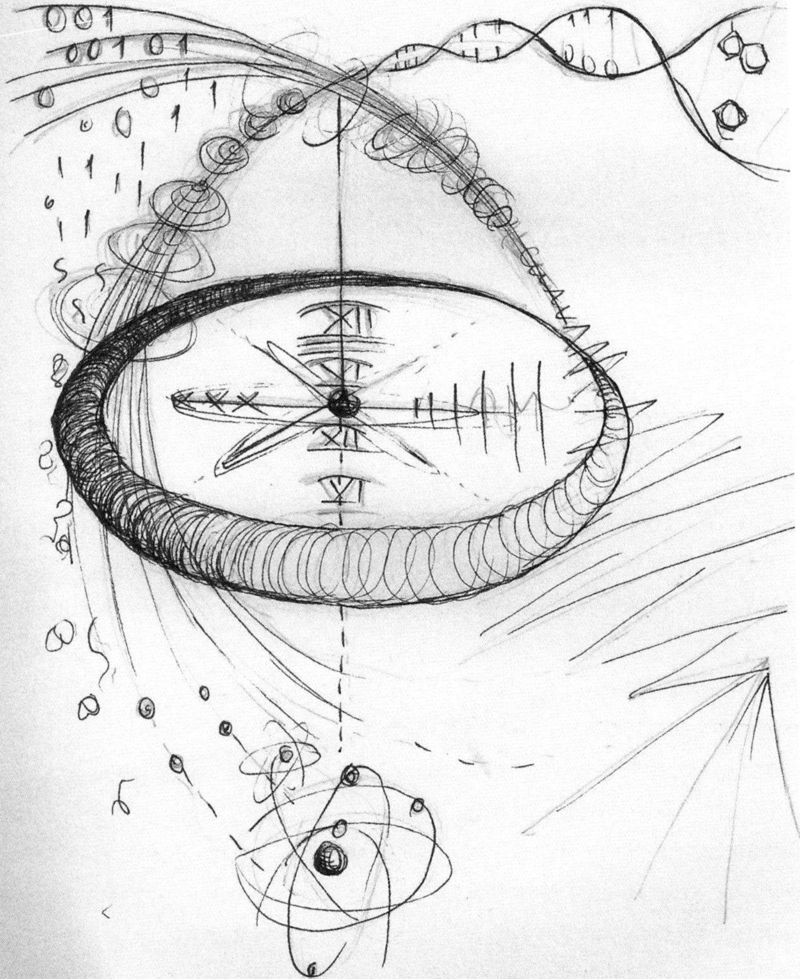

other fundamental forces as well, a strong force that binds together atomic nuclei and a weak force that governs radioactive decay, not to mention a whole zoo of particles such as quarks, muons, gluons and neutrinos. Most other physicists seemed unconcerned about formulating a theory that would unite electromagnetism and gravity. The focus was on the strange and wonderful new world of quantum mechanics that had just been opened up for exploration.

But Einstein wasn't completely alone in his quest; several other scientists turned their attention to the problem of unification as well. In 1918, Hermann Weyl had proposed a unification scheme based on a generalization of the geometry of curved space that Einstein had used in developing his general theory of relativity. Inspired by Weyl's work, Theodor Kaluza showed that if spacetime were extended to five dimensions, then four of those dimensions would encompass Einstein's general relativity equations, while the fifth dimension would be the equivalent for Maxwell's equations for electromagnetism. Oskar Klein later determined that the fifth dimension would be curled up so small that we can't detect it.

Einstein was quite taken by the five-dimensional approach. In 1919, he wrote to Kaluza, 'The idea of achieving unification by means of a five-dimensional cylinder world would never have dawned on me… At first glance I like your idea enormously.' Einstein also explored the approach of extending general relativity to include electromagnetism but still keeping the four-dimensional geometry of spacetime. He persisted with both approaches for the last 30 years of his life, but he never found the answers he was looking for. 'Most of my intellectual offspring end up very young in the graveyard of disappointed hopes,' he lamented in 1938.

Einstein spent the last decades of his life refining his ideas on a unified theory, while also trying to solve what he saw as problems in his general theory of relativity, such as its prediction of black holes, which he simply didn't like.

One unfortunate consequence of Einstein's search for unification was that it, to some extent, isolated him from the rest of the physics community. There were many physicists who thought that Einstein contributed nothing of importance to the field in the last 20 years of his life. In particular, Einstein's antipathy towards quantum mechanics may have cut him off from some promising lines of research. It was in large part his belief that quantum mechanics was flawed that spurred Einstein on in his unification quest. He believed beyond doubt in a Universe that exists completely independently and doesn't wait until it is observed to take form as quantum mechanics would have it. 'Do you really believe that the Moon is not there unless we are looking at it?' he asked.

Einstein was aware of this shortcoming in his dealings with quantum mechanics, commenting in 1954 that, 'I must seem like an

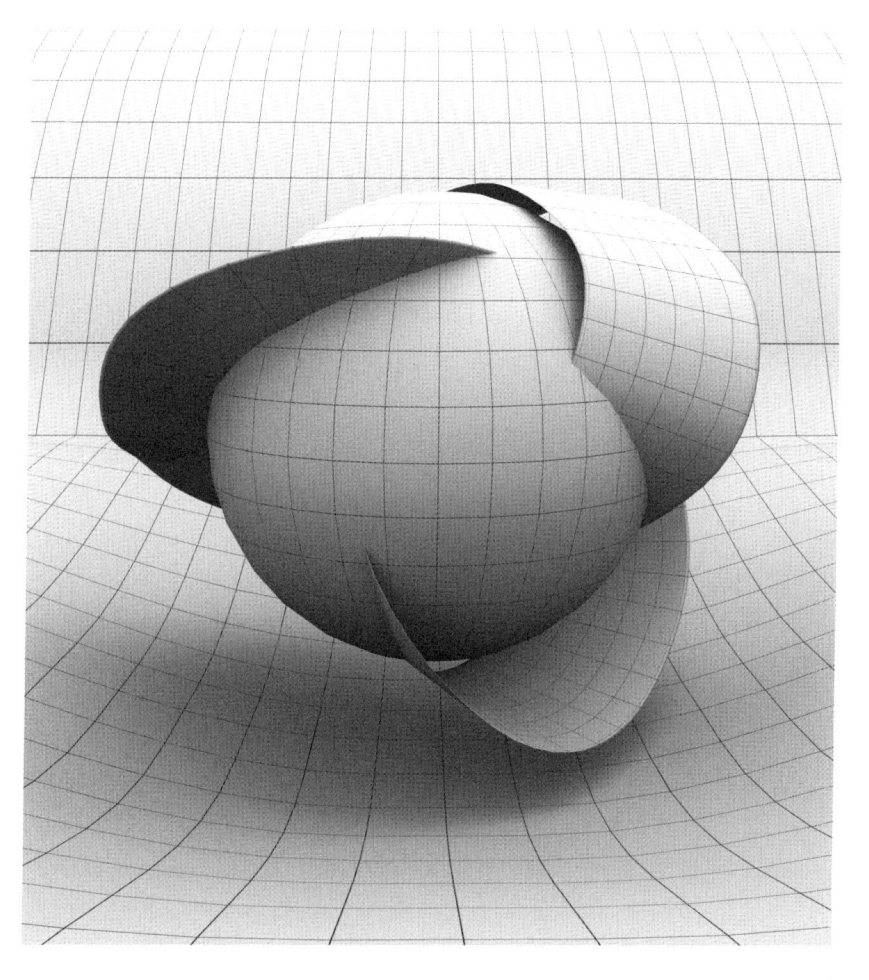

Can we find a theory of everything? | **293**

ostrich who forever buries its head in the relativistic sand in order not to face the evil quanta.' Nearing the end of his life, and perhaps realizing that his quest was going to be a fruitless one, he wrote: 'I have locked myself into quite hopeless scientific problems, the more so since, as an elderly man, I have remained estranged from the society here.'

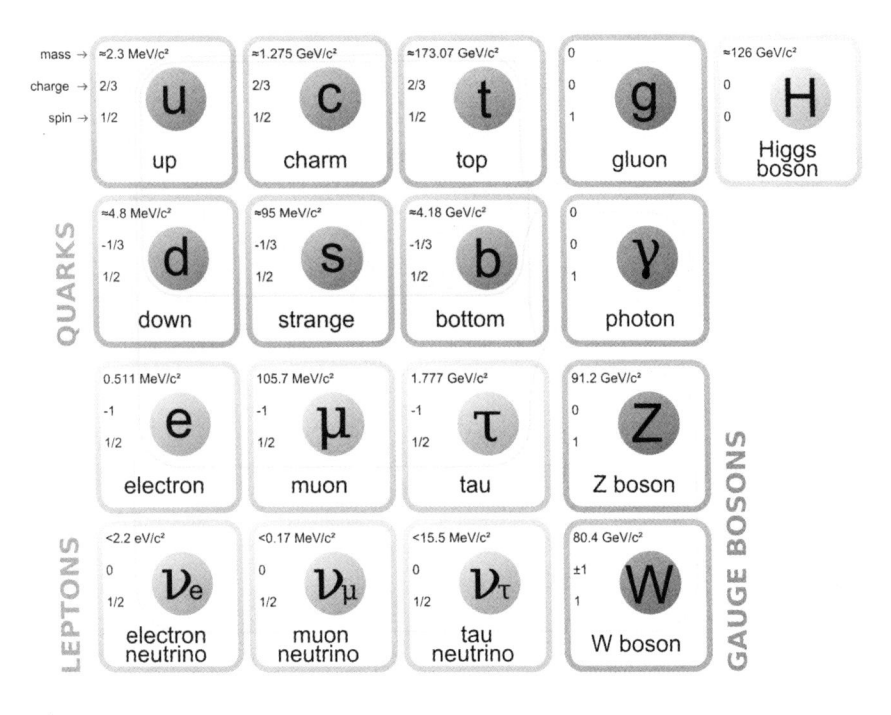

	mass →	charge →	spin →		name
QUARKS	≈2.3 MeV/c^2	2/3	1/2	u	up
	≈1.275 GeV/c^2	2/3	1/2	c	charm
	≈173.07 GeV/c^2	2/3	1/2	t	top
	≈4.8 MeV/c^2	-1/3	1/2	d	down
	≈95 MeV/c^2	-1/3	1/2	s	strange
	≈4.18 GeV/c^2	-1/3	1/2	b	bottom
LEPTONS	0.511 MeV/c^2	-1	1/2	e	electron
	105.7 MeV/c^2	-1	1/2	μ	muon
	1.777 GeV/c^2	-1	1/2	τ	tau
	<2.2 eV/c^2	0	1/2	ν_e	electron neutrino
	<0.17 MeV/c^2	0	1/2	ν_μ	muon neutrino
	<15.5 MeV/c^2	0	1/2	ν_τ	tau neutrino
GAUGE BOSONS	0	0	1	g	gluon
	0	0	1	γ	photon
	91.2 GeV/c^2	0	1	Z	Z boson
	80.4 GeV/c^2	±1	1	W	W boson
	≈126 GeV/c^2	0	0	H	Higgs boson

Perhaps Einstein was simply ahead of his time. Decades after his death the search for a 'theory of everything' has become a holy grail for many physicists.

The Standard Model

During the 1960s and 1970s, particle physicists discovered that there are two basic building blocks of matter – they are the fundamental particles known as quarks and leptons. Quarks are always found within larger particles, such as protons and neutrons; they are never found separately in nature and are always bound with other quarks, held together by the short-range strong nuclear force. The leptons, which include the electron, are not affected by the strong force. However, both quarks and leptons are affected by the weak nuclear force, which is responsible for certain types of radioactivity. The strong force is stronger than the electromagnetic force over distances smaller than an atom while the weak force is feebler than either. Gravity is the weakest of the four fundamental forces but it acts over infinite distances. The electromagnetic force is much stronger than gravity and also has an infinite range.

In 1968, Sheldon Glashow, Steven Weinberg and Abdus Salam announced a unified theory of electromagnetism and the weak nuclear force. Their electroweak theory, as it was called, suggested that the weak force was carried by particles called W and Z bosons.

These were subsequently discovered in the 1980s.

Today, physicists believe that in the early Universe, shortly after the Big Bang, the strong and weak nuclear and electromagnetic forces were unified. All three of these fundamental forces result from the exchange of force-carrier particles, like the W and Z bosons. Each fundamental force has its own corresponding boson – the strong force is carried by the gluon, and the electromagnetic force is carried by the photon. The Standard Model of particle physics, as it is called, was developed in the early 1970s and explains how the electromagnetic, strong and weak forces, along with all their associated carrier particles, act on all of the matter particles. It works very well, but it has shortcomings. There still seems no way to merge gravity with quantum mechanics and gravity is not part of the Standard Model. On the scale at which particle physics operates, the effect of gravity is so weak as to be negligible, which means that its exclusion has no effect on the predictions of the Standard Model.

Elementary Particles

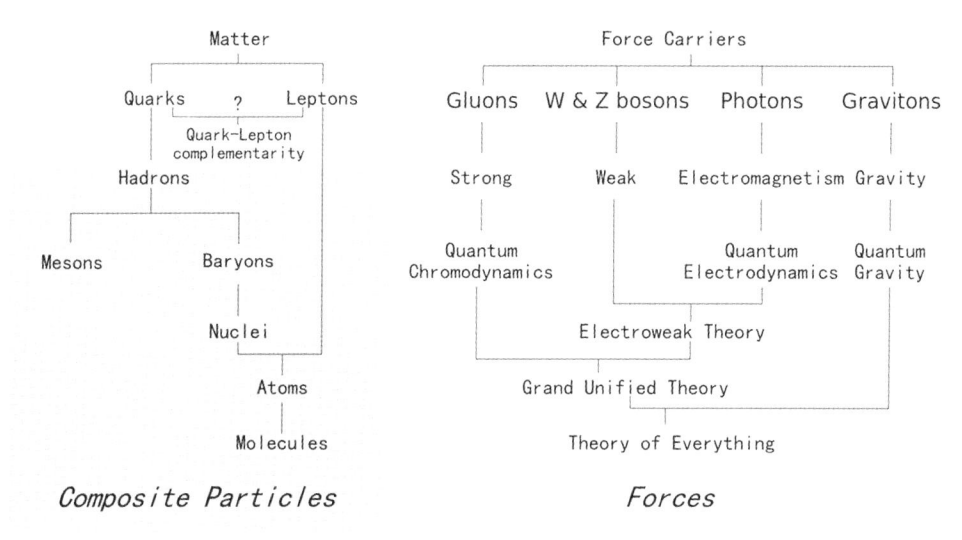

```
          Matter                                Force Carriers
     ┌──────┴──────┐                 ┌──────┬──────┴──────┬──────┐
   Quarks    ?   Leptons          Gluons  W & Z bosons  Photons  Gravitons
     └─────┬─────┘                   │         │           │         │
       Quark-Lepton                  │         │           │         │
       complementarity               │         │           │         │
          │                        Strong    Weak   Electromagnetism Gravity
       Hadrons                       │         │           │         │
     ┌────┴────┐                     │         │           │         │
  Mesons    Baryons                Quantum            Quantum     Quantum
               │              Chromodynamics      Electrodynamics  Gravity
             Nuclei                   │         │           │         │
               │                      │      Electroweak Theory       │
             Atoms                    │             │                  │
               │                      └─── Grand Unified Theory ──────┘
           Molecules                              │
                                        Theory of Everything
```

Composite Particles   *Forces*

String theory

Currently, one of the most hopeful candidates for a theory of everything is string theory. Not only does it promise a theory of gravity on the microscopic scale, it also seeks to provide a unified and consistent description of the fundamental structure of the Universe, uniting all four fundamental forces and the fundamental particles of the Standard Model.

MATTER

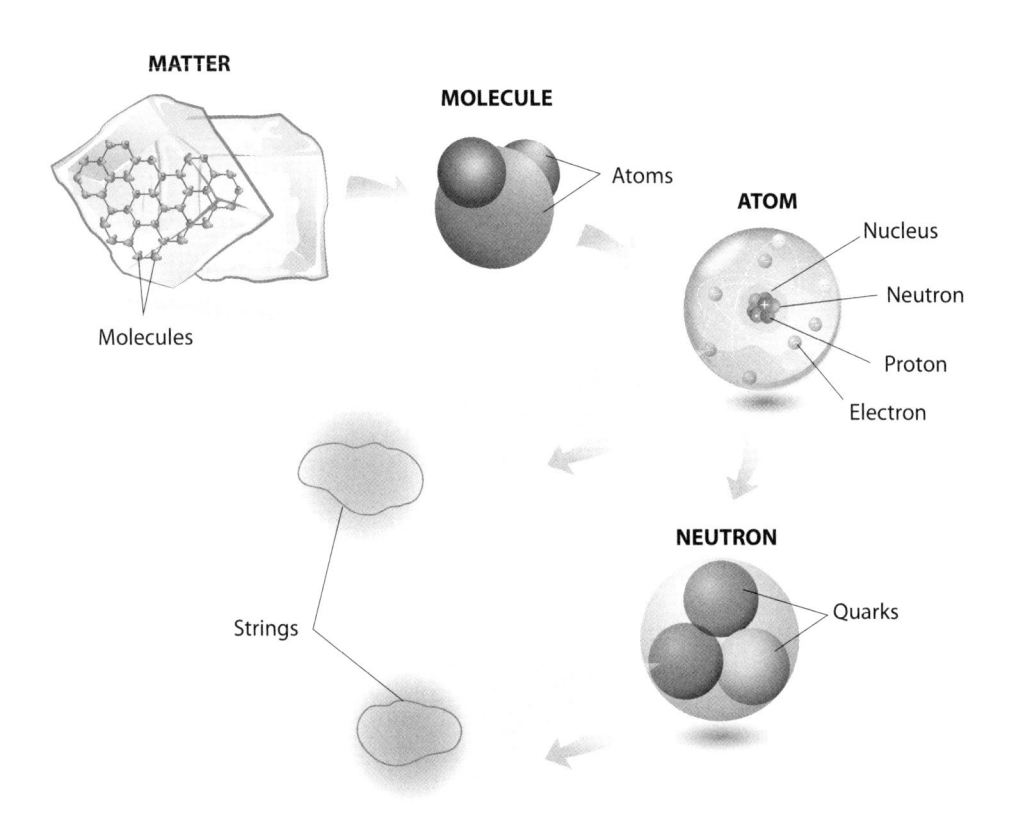

MOLECULE

Molecules

Atoms

ATOM

Nucleus

Neutron

Proton

Electron

NEUTRON

Quarks

Strings

In December 1984, John Schwarz, of the California Institute of Technology, in Pasadena, and Michael Green, of Queen Mary College, London University, published a paper showing that string theory could build a bridge across the mathematical chasm that separates general relativity and quantum mechanics.

At the heart of string theory is the idea that all of the different fundamental particles are really just different manifestations of one basic object: a string. Since the early 20th century, nature's fundamental particles, such as electrons, quarks and neutrinos, have been portrayed as being as-small-as-you-can-go objects with no internal structure. String theory challenges this. It proposes that at the heart of every particle is a tiny, vibrating string-like filament. The differences between one particle and another – their mass, charges and other properties – all depend on the vibrations of their internal strings. Like a skilled violinist conjuring a melody, nature manifests all the particles of the atomic realm through changes in the frequency of a one-dimensional subatomic string.

Of great interest is the fact that one of the 'notes' on the string corresponds to the graviton. The graviton is a hypothetical particle that, according to quantum physics, should carry the force of gravity from one location to another, just as the photon does for the electromagnetic force. This seemed to promise a way to get gravity and quantum mechanics working together.

So, are strings 'real'? Could we go looking for them in CERN's Large Hadron Collider, for example? Unfortunately, that just isn't possible. The mathematics of string theory require them to be about a million

> **WHAT IS GRAVITY – CURVED SPACETIME OR GRAVITONS?**
>
> Theorists currently believe that descriptions of gravity as being the result of the curvature of spacetime brought about by the matter in it, or as an exchange of graviton force particles, are both equally valid; just as we can think of electromagnetism as being either the result of changes in the electromagnetic field or as an exchange of photons. The problem is that, although, thanks to Einstein and the general theory, physicists have a workable solution for gravity that involves the gravitational field, and gravitational forces as a curvature of spacetime, there is no current quantum theory of gravity involving gravitons that is as well worked out and proven by experiment as Einstein's. String theory suggests that gravitons could exist but, as yet, there is no experimental proof of their existence.

billion times smaller than anything the world's most powerful particle accelerators have uncovered. Unless, as physicist Brian Greene says, we can build 'a collider the size of the galaxy', we have no hope of directly detecting strings.

An additional complication of string theory is that its equations require that the Universe has extra spatial dimensions to make them work. String theorists took up the idea first developed by Kaluza and Klein in the early years of the 20th century when they were trying to link Einstein's gravity with electromagnetism. Perhaps,

Can we find a theory of everything? | 301

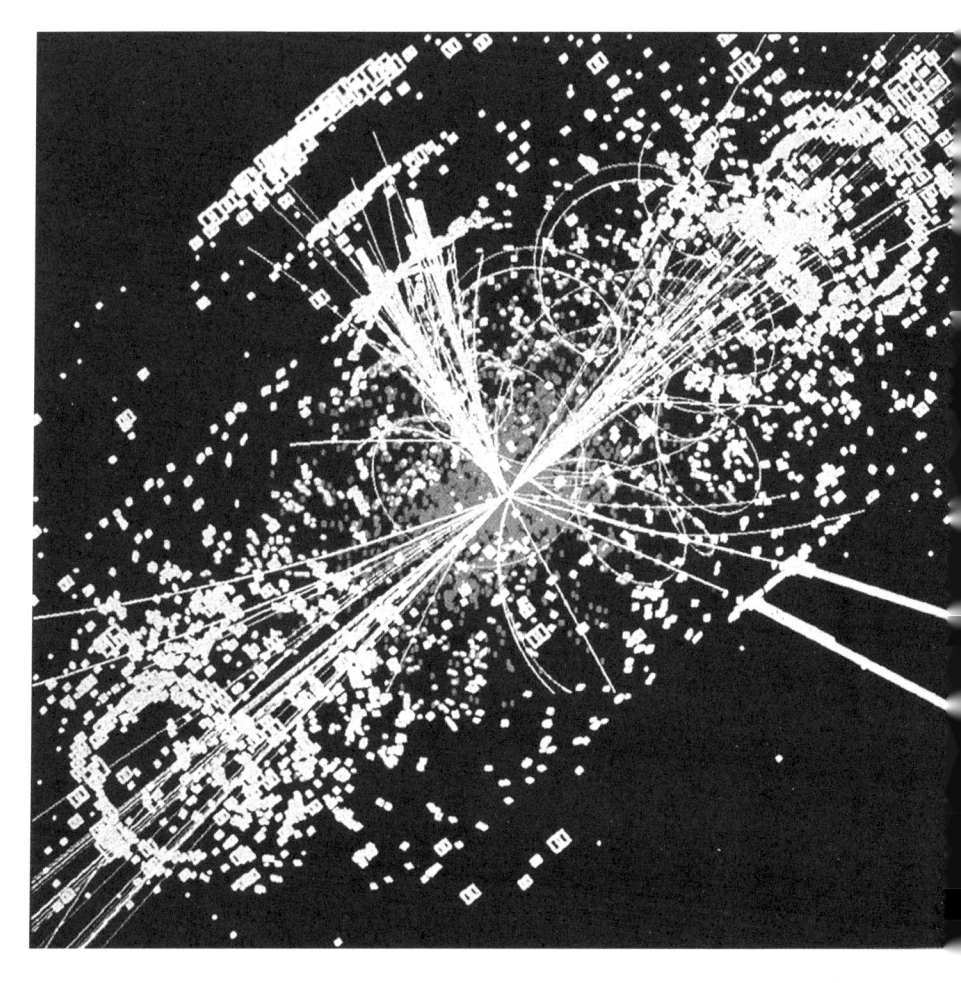

they suggested, the Universe has the three big dimensions that we all know and move through, but there might also be additional dimensions that were so tiny and compact, wound up inside the 'normal', that they were beyond our ability to detect.

A group of theorists suggested that because strings are so small they will vibrate not just in the 'big' dimensions, but in the tiny ones as well. Audaciously, they predicted that since it is the vibrations that determine the properties of the fundamental particles, which we can detect experimentally, and the vibrations are determined by the shape of the extra dimensions, there might be a way to work back to a map of the esoteric dimensions.

Unfortunately, it appeared that the number of mathematically allowable shapes for the extra dimensions ran into the billions. Theorist Leonard Susskind suggested that, if there was no one shape that was right then maybe they all were. Perhaps all of the shapes are the right shape within their own unique Universe. Our Universe would just be one of a vast, perhaps infinite assembly, each with features that are determined by the shape of their extra dimensions. 'Our' Universe's hidden dimensions make possible the laws of physics that led to the existence of stars and galaxies, the chemical elements, life itself. In some other dimensional configuration different laws would apply and the Universe would be a very different, probably lifeless place.

These heady ideas mirror developments in cosmology that have suggested that the Big Bang may not have been a unique event. Instead, so the theory goes, there have been infinite bangs giving rise to an infinity of expanding universes, called the multiverse, perhaps if Susskind is right, each with its unique complement of compact dimensions.

Can any of this be true? In theory, yes it could, though we could never have any way of knowing for sure. But as we have seen, the history of science demonstrates that we shouldn't dismiss ideas out of hand just because they go against what 'common sense' suggests we should expect. With that sort of attitude we wouldn't have embraced quantum physics, or Einstein and his relativity theories.

What would Einstein have made of it all? Would he have embraced the mathematics of string theory or would he have found some flaw in it and reject it as he did with quantum mechanics? Would the man who was convinced that 'God does not play dice' accept the notion of an infinity of universes, the characteristics of which were each determined by a fresh roll of the cosmic dice? It's most likely that he would be fascinated by it all.

In a letter he wrote in 1953, Einstein explained: ***'Every individual... has to retain his way of thinking if he does not want to get lost in the maze of possibilities. However, nobody is sure of having taken the right road, me the least.'***

Picture Credits

Public domain: 6, 13, 27, 30, 34, 35 (credit to Sascha Grusche), 42 (credit to Impensustained), 47 (credit to AetherWind), 48, 53, 54, 65, 66 (credit to Douglas Hofstadter), 68, 72, 76 (credit to ZEISS Microscopy), 79 (credit to Wellcome Images), 80, 81, 82 (credit to Nein Arimasen), 83, 84, 87, 91 (credit to Michael Schmid), 92 (credit to Santi Villamarin), 97, 97, 99 (credit to Mats Lindh), 102, 104, 107, 108, 109, 110, 111, 113 (credit to Wolfgangbeyer), 155, 116 (credit to Lutfar Rahman Nirjhar), 117 (credit to P. Fraundorf), 118, 123, 124, 126 (No-w-ay in collaboration with H. Caps), 127, 129 (MichaelMaggs Edit by Richard Bartz), 130 (Lienhard Schulz), 133, 134, 137, 139 (Saffron Blaze), 140, 146, 147 (Markus Poessel [Mapos]), 155 (Lars H. Rohwedder, Sarregouset), 158, 161, 163, 165, 167, 168, 169, 171, 175, 177, 178 (Andrew Dunn), 180, 181, 183 (Adam Baker), 185, 188, 194, 196 (Hubble ESA), 197 (Adam Evans), 198, 199 (MGrundy), 203, 204, 211 (K. "bird" N.), 213 (Brews Ohare), 214, 215, 216, 219 (NASA Goddard Space Flight Center from Greenbelt, MD, USA), 226 (NASA), 228 (Anne-Lise Heinrichs), 229, 231, 233 (Biswarup Ganguly), 234 (Cmglee), 237, 239, 240, 241 (Szdori), 242, 243, 246 (Richard Parsons), 247 (Friedrich Hund), 248 (Berndthaller), 249, 251, 252, 257, 260, 261, 264 (Peng), 266, 276 (Jeff Dahl), 277, 283, 284, 286, 287, 294, 296 (Molendijk-Bart - Anefo), 297 (Headbomb), 301 (Lucas Taylor_CERN)

Science Photo Library: 105, 144, 152, 166, 205, 207, 222, 225, 263, 273, 280, 282

Shutterstock: 4, 7, 9, 17, 20, 22, 23, 25, 29, 31, 32, 36, 38, 40, 44, 45, 50, 51, 52, 55, 59, 60, 64, 67, 70, 71, 73, 77, 90, 94, 96, 100, 121, 125, 135, 142, 149, 150, 156, 162, 173, 176, 182, 186, 195, 201, 208, 209, 212, 220, 221, 236, 238, 245, 258, 271, 274, 279, 288, 290, 293, 298